FOREWORD

The OECD/NEA Nuclear Science Committee set up a Working Party on Physics of Plutonium Recycling in June 1992 to deal with the status and trends of physics issues related to plutonium recycling with respect to both the back end of the fuel cycle and the optimal utilisation of plutonium. For completeness, issues related to the use of the uranium coming from recycling are also addressed.

The Working Party met three times and the results of the studies carried out have been consolidated in the series of reports "Physics of Plutonium Recycling".

The series covers the following aspects:

- Volume I *Issues and Perspectives*;
- Volume II *Plutonium Recycling in Pressurized-Water Reactors*;
- Volume III *Void Reactivity Effect in Pressurized-Water Reactors*;
- Volume IV *Fast Plutonium-Burner Reactors: Beginning of Life*;
- Volume V *Plutonium Recycling in Fast Reactors*; and,
- Volume VI *Multiple Recycling in Advanced Pressurized-Water Reactors.*

The present volume is the second in the series and describes the specific benchmark studies concerned with the calculation of physics parameters of a pressurized-water reactor fuelled with plutonium from different recycles.

The opinions expressed in this report are those of the authors only and do not represent the position of any Member country or international organisation. This report is published on the responsibility of the Secretary-General of the OECD.

CONTENTS

SUMMARY

The report presents the main results of benchmark calculations performed for two different MOX pin cells of PWR type representing typical fuel for plutonium recycling. The benchmark was defined by the Nuclear Science Committee of the OECD Nuclear Energy Agency and 12 institutions from 9 different countries contributed 14 solutions for which different methods and basic nuclear data are used. One-group reaction rates, cross-sections and number densities of 17 actinides and 21 fission products and neutron spectra are compiled. Reaction rates and number densities are presented in tables and partially in plots to enable detailed analysis.

CONTRIBUTORS

AUTHORS	*D. Lutz*	IKE	Germany
	W. Bernnat	IKE	Germany
	K. Hesketh	BNFL	U.K.
	E. Sartori	OECD/NEA	
PROBLEM SPECIFICATION	*H. Küsters*	KFK	Germany
	G. Schlosser	Siemens	Germany
	J. Vergnes	EDF	France
	H. W. Wiese	KFK	Germany
DATA COMPILATION AND ANALYSIS	*D. Lutz*	IKE	Germany
	A. & W. Bernnat	IKE	Germany
SUPPLEMENTARY ANALYSIS	*S. Cathalau*	CEA	France
	A. Puill	CEA	France
	H. Takano	JAERI	Japan
	E. Saji	Toden	Japan
TEXT PROCESSING AND OUTLAY	*P. Jewkes*	OECD/NEA	

BENCHMARK PARTICIPANTS

Th. Maldague	Belgonucléaire	Belgium
P. Marimbeau	CEA	France
A. Puill	CEA	France
P. Barbrault	EDF	France
J. Vergnes	EDF	France
A. Kolmayer	Framatome	France
W. Bernnat	IKE	Germany
S. Käfer	IKE	Germany
D. Lutz	IKE	Germany
M. Mattes	IKE	Germany
W. Hetzelt	Siemens	Germany
G. Schlosser	Siemens	Germany
K. Ishii	Hitachi Ltd.	Japan
H. Maruyama	Hitachi Ltd.	Japan
H. Akie	JAERI	Japan
H. Takano	JAERI	Japan
J. M. Li	ECN	Netherlands
V. A. Wichers	ECN	Netherlands
K. Ekberg	Studsvik	Sweden
F. Holzgrewe	PSI	Switzerland
J. M. Paratte	PSI	Switzerland
G. Mangham	BNFL	U.K.
R. N. Blomquist	ANL	U.S.A.

Introduction

The recycling of plutonium in PWRs in the form of mixed oxide (MOX) uranium plutonium fuel assemblies is a technology which is now well established and many countries have many years' experience to draw on. Within the constraints of current fuel management schemes, discharge burnups and plutonium isotopic vectors, it is fair to say that physics methods are available which can be considered to be mature and fully proven.

The validity of present methods cannot be assumed to extend outside the current constraints, however, and further validation will be required to demonstrate that both the basic nuclear data and the calculational methods remain adequate for the more challenging problems that are expected to arise within the next decade. The challenges to existing physics methods will come from high burnup fuel management schemes and feed plutonium with lower fractions of the fissile isotopes Pu-239 and Pu-241. The effect of both these changes will be to increase the total plutonium loading necessary in the MOX fuel. This will increase the thermal neutron absorption and drastically alter the thermal neutron spectrum.

Unfortunately, experimental validation will not be forthcoming for this new situation for several years; yet it is important to have some indication of what level of development effort will be required to address the possible shortcomings of present physics methods. Faced with this situation, the WPPR committee agreed that a set of benchmark exercises would be a valuable means of making progress in the interim period before any practical results become available from in-reactor irradiation experience. It was hoped that a comparison of the results would give valuable insights into the likely requirements as regards improving the nuclear data and methods. While accepting that such benchmarks could not possibly identify the 'true' answer, it was anticipated that a consensus view on the most probable answers would emerge which would be helpful in guiding future work.

Objectives of the benchmarks

Two benchmarks were devised for MOX in PWRs. They are simple infinite array pin cell problems designed to allow intercomparison of infinite multiplication factors as a function of burnup.

- The first such pin cell problem, designated '*Benchmark A*' comprises a pin cell with plutonium of a low isotopic quality (i.e., a low fraction of the thermally fissile isotopes Pu-239 and Pu-241). It is expected that such plutonium will become available for recycling at some future date when MOX fuel assemblies are themselves reprocessed. The quality of plutonium recovered from PWR spent fuel decreases during each recycle, the rate depending on the discharge burnup of the reactor fuel cycle and on the ratio in which MOX assemblies are blended with UO_2 assemblies in the reprocessing plant. The particular isotopic composition specified for Benchmark A represents a hypothetical case of the fifth recycle of plutonium for a scenario in which MOX assemblies are blended with UO_2 assemblies in a ratio which reflects that which will arise in a self-generation recycle mode in a PWR. The total plutonium content is 12.5 w/o (6.0 w/o fissile) and the isotopic vector is as follows:

Pu-238	Pu-239	Pu-240	Pu-241	Pu-242
4 %	36 %	28 %	12 %	20 %

The poor plutonium isotopic quality in Benchmark A demands a high concentration of total plutonium in order to compensate for neutron absorption in Pu-240 and Pu-242 isotopes. The high plutonium concentration poses a severe challenge to existing nuclear data libraries and lattice codes, which was the driving force behind the specification.

- The other pin cell problem, designated '*Benchmark B*', specified a plutonium isotopic vector with a higher fissile fraction that is representative of commercial PWR MOX recycle at the present time. The total plutonium content is 4.0 w/o (2.8 w/o) fissile with the following isotopic vector:

Pu-238	Pu-239	Pu-240	Pu-241	Pu-242
1.8 %	59 %	23 %	12.2 %	4.0 %

This problem was intended to act in the form of a 'control' to show whether the spread of results in the more challenging problem could be attributed to the poor plutonium vector or to underlying differences in the nuclear data and methods, which also apply to today's situation.

The full specifications of Benchmarks A and B can be found in Appendix A.

Participants, methods and data

A total of 14 solutions were contributed for Benchmark A and 13 for Benchmark B, representing 12 institutions from 9 countries. A full list of all the contributors is provided below. This list identifies the codes and nuclear data libraries used by the various contributors and where necessary makes pertinent remarks. The letters in parentheses give the abbreviations which will be used to identify each contributor throughout this report. Table 1 summarises the same information.

1. ***Argonne National Laboratory, (ANL), U.S.A.***

Participant: R. N. Blomquist
Code: VIM (continuous Monte Carlo)
Data Library: ENDF/B-V
Remarks: Participants can compare these results against their own by carrying out one additional calculation at 300 K. For more information see Appendix B.1

2. ***Belgonucléaire (BEN), Belgium***
   Participant: Th. Maldague
   Code: LWRWIMS
   Data Library: 1986 WIMS

3. ***British Nuclear Fuels (BNFL), U.K.***
   Participant: G. Mangham
   Code: LWRWIMS
   Data Library: 1986 WIMS
   Remarks: Detailed information is available in Report FEDR 93/2050 [1]

4. ***Commissariat a l'Energie Atomique (CEA) and Framatome, France***
   Participant: A. Puill and A. Kolmayer
   Code: APOLLO 2
   Data Library: JEF-2.2, CEA-93
   Remarks: Details are in Appendix B.2

5. ***ECN Nuclear Energy (ECN), Netherlands***
   Participant: V. A. Wichers and J. M. Li
   Code: SCALE 4 and WIMS-D
   Data Library: JEF-2.2 and SCALE
   Remarks: Details are in Appendix B.3

6. ***Electricité de France (EDF), France***
   Participant: P. Marimbeau (CEA), P. Barbrault, J. Vergnes
   Code: APOLLO 1
   Data Library: CEA-86
   Remarks: Details are in Appendix B.4

7. ***Hitachi Ltd. (HIT), Japan***
   Participant: K. Ishii and H. Maruyama
   Code: VMONT
   Data Library: JENDL-2, ENDF/B-IV
   Remarks: Details are in Appendix B.5

8. ***University of Stuttgart (IKE-1), Germany***
   Participant: D. Lutz
   Code: CGM, RSYST
   Data Library: JEF-1
   Remarks: Details are in Appendix B.6

9. ***University of Stuttgart (IKE-2), Germany***
   Participant: W. Bernnat, M. Mattes, S. Käfer
   Code: MCNP 4.2
   Data Library: JEF-2.2
   Remarks: Point data recalculated with NJOY 91.91 for the temperatures 300 K and 600 K. Details are in Appendix B.7

10. ***Japan Atomic Energy Research Institute (JAE), Japan***

Participant: H. Akie and H. Takano
Code: SRAC
Data Library: JENDL-3.1
Remarks: Details are in Appendix B.8

11. ***Paul Scherrer Institut (PSI-1), Switzerland***

Participant: J. M. Paratte
Code: BOXER
Data Library: JEF-1
Remarks: Details are in Appendix B.9

12. ***Paul Scherrer Institut (PSI-2), Switzerland***

Participant: F. Holzgrewe
Code: CASMO 3
Data Library: ENDF/B-IV
Remarks: Details are in Appendix B.10

13. ***Siemens (SIE-1), Germany***

Participant: G. Schlosser, W. Hetzelt
Code: CASMO 3
Data Library: J70

14. ***Studsvik Core Analysis (STU), Sweden***

Participant: K. Ekberg
Code: CASMO 4
Data Library: JEF-2.2
Remarks: Details are in Appendix B.11

The two CASMO 3 solutions (12 and 13) were withdrawn.

As can be seen most conveniently from Table 1, most of the contributors to Benchmarks A and B used deterministic lattice codes. These are the usual tools used for nuclear design applications such as calculating reactivities and irradiation depletion effects. Two contributors used Monte Carlo methods, which provide a useful cross-check on the methods, but which cannot carry out depletion calculations and are therefore restricted to the zero burnup step. The Monte Carlo codes are also restricted in that nuclear data tabulations are only usually available for a limited set of materials and temperatures. Table 1 highlights where the temperatures available did not coincide with the benchmark specifications.

At this stage it is appropriate to draw attention to some of the special physics aspects that need to be accounted for in the MOX benchmark calculations and to highlight the aspects which participants took particular care to model rigorously:

1. The relatively large thermal absorption cross-sections of plutonium considerably reduces the thermal neutron flux compared with uranium, while the flux at higher energies is less drastically affected. The result is that the neutron spectrum in a MOX assembly is much harder than that in a UO_2 assembly and the resolved resonances have a much higher impact on the calculation of group cross-sections.

2. In addition, the unresolved resonances and the threshold reactions in the MeV range also require more careful attention. Some of the contributors used codes where resonance self-shielding in all plutonium isotopes is treated rigorously, and this has an important bearing on the results, as will be seen later.

Table 2 provides information concerning the resonance treatments used by the various participants in Benchmarks A and B. Most applied f-factors (Bondarenko) to allow for resonance self-shielding, while some performed ultra fine cell calculations to account for both mutual shielding and local effects. Appendices B provide more detailed information.

The energy per fission values to be used were defined in the benchmark specifications with five isotopes only contributing to energy release. Table 3 indicates that six of the participants used the specified values. While the EDF and CEA solutions omitted according to the specification the energy release from other isotopes applying a specifically prepared library, the other participants calculated the energy production according to their normal design methods, which account for all fissile contributions. The effect is that the EDF and CEA solutions have slightly stretched effective burnup scales compared with the other solutions.

Most participants took account of (n,2n)-reactions by lowering the absorption cross-sections artificially. The effect increases the multiplication factor by about 0.2%. A rigorous treatment, however, involves modifying the actinide chains explicitly and shows consequently that artificially reducing the absorption cross-sections introduces a systematic error due to the higher levels of Np-237 which build up, for higher burnups.

The influence of the fission spectrum being inappropriate to the actual fuel composition is of the same order of magnitude.

Results

Figures 1-A and 1-B show compilations of k-infinities for Benchmarks A and B respectively from the various participants. Tables 4-A and 4-B list the same data. These are the principal results of the benchmarks. The spread of results at zero irradiation is 3.1% for Benchmark A and 1.3% for Benchmark B. There is also some spread in the slope of k-infinities versus burnup. This is more clearly seen in Table 5-A and 5-B, which show the reactivity changes versus burnup, which vary from 15% to 18%. Tables 6-A and 6-B show the one-group fluxes as a function of burnup.

For the discussion of k-infinity only the reaction rates and ν values are necessary. An overview of spread of these functions is presented in Tables 7 to 9.

Fission and absorption rates have been renormalised where necessary to total absorption rate in the cell equal to 1 for easy comparison. The deviation of those normalised rates from the best estimate values are a direct measure of the deviations of the corresponding multiplication factors. The applied normalisation procedure neglects the (n,2n)-effect.

Detailed information about 17 actinides and 21 fission products are also presented. The selection has been made on the base of the absorption rates of the JAERI results for Benchmark A.

Tables 11 to 16 [1] show the absorption rates of actinides and fission products for Benchmark A, the fission rates and ν values and the number densities of actinides and fission products, respectively. The corresponding functions for Benchmark B are presented in Tables 17 to 22 [1] . Graphical presentation of absorption and fission rates of the actinides is provided in Appendix C. The pages are labelled with A-ar and B-ar for absorption and A-fr and B-fr for fission rates, respectively.

The last type of results are the burnup-dependent spectra. They have been normalised (total energy integral = 1.0) to enable comparisons between different burnup states and also different contributions. For each contribution two figures are given, the spectrum per lethargy for fresh fuel in a logarithmic scale, and its modifications during the burnup in a linear scale.

For better understanding of discrepancies it is helpful to compare also the cross-sections and number densities for sensitive nuclides. Therefore the corresponding tables and plots are provided, a part of it is included in this report. The complete information is available in computer readable form from the NEA Data Bank as postscript files with self explaining names. The abbreviations used are as follows:

- **ar** absorption rates,
- **fr** fission rates,
- **nu** number of neutrons per fission,
- **td** nuclide number densities,
- **sa** microscopic one group absorption cross section,
- **sf** microscopic one group fission cross section.

The files start with the information relative to the actinides ordered according to the charge and the atomic weight number and continue with the data of fission product isotopes.

Discussion

Multiplication factors

Referring to Figure 1-A, a disappointingly large spread of k-infinities for Benchmark A (approaching 3.1% at zero burnup) can be observed; it is encouraging though that there is a substantial agreement as to the slope of k-infinity with burnup. Some of this spread is, however, straightforward to account for.

Not all current lattice codes are able to treat accurately resonance absorption in the higher plutonium isotopes. This is because historically the absolute concentrations of the higher plutonium isotopes in both UO_2 and MOX fuels have always been low enough that self-shielding in them could safely be neglected. As explained earlier, the purpose of Benchmark A was to test code predictions in a challenging situation where this no longer applies. Thus Benchmark A specifies 3 w/o absolute of Pu-242, for which self-shielding can by no means be neglected. In view of this, it is not surprising that some of the results are systematically in error. For the conditions of Benchmark A, the effect is estimated to be worth a systematic bias of about 2.5% in k-infinity, so that the code predictions in which higher isotope self-shielding is not applied, should be increased by this amount. The solutions provided by BEN and BNFL (both LWRWIMS) fall into this category. From Figure 1-A, it is apparent that if these contributions are corrected upwards by 2.5%, or if only those codes with rigorous higher isotope

[1] Densities are given in $10^{24}/cm^3$ in Tables 15, 16, 21 and 22.

self-shielding are included, the spread of results is considerably narrowed to about 0.9 to 1.5%, depending on the burnup.

Considering the solutions incorporating rigorous self-shielding, the 0.9% spread in k-infinities most probably arises from underlying differences in the nuclear data libraries or different methods applied for taking into account the resonance shielding effects (compare Table 2 and Appendix C).

A special benchmark was established during the WPPR meeting in November 1994 in order to quantify the portion of these differences due to applied physics methods. For the fresh state only of the pin cell of Benchmark A and B in square geometry results are being computed applying the data bases JEF-2.2, JENDL-3.1 and JENDL-3.2. The results will be published and analysed separately.

There is a clear tendency for solutions based on a common data library to be very close, e.g., PSI 1 and IKE 1 (both using JEF-1) as one sub-group, CEA, ECN, IKE 2 and STU (all using JEF-2.2) as a second sub-group and HIT and JAE (both using JENDL-3.1) as the third one. This suggests that differences in the lattice code methods are less important than the nuclear data evaluations.

In respect of the 1.5% residual spread, it has to be said that if this was representative of the uncertainty on the lattice calculations, it would be unacceptable for design and licensing applications. Current nuclear design methods typically claim uncertainties on reactivity of about 0.2% with occasional outliers of up to 0.5%. A concerted effort will clearly be necessary to resolve the outstanding differences and this will necessitate experimental validation. The situation is particularly unsatisfactory because the reactivity of MOX fuel tends to increase only very slightly as the plutonium content increases, an effect which is greatly exaggerated in the Benchmark A situation because of the low fissile fraction of plutonium. Thus, any attempt to increase reactivity by loading a higher fissile plutonium content is to a large extent opposed by the increased absorption from the even isotopes. This means that any uncertainty in the reactivity predictions will translate into a disproportionately large spread in the plutonium concentration needed to achieve the specified lifetime reactivity.

The codes and libraries give a better agreement for the more conventional MOX fuel of Benchmark B and the results are closer (see Figure 1-B). The same grouping of solutions as in Benchmark A is also visible in Table 4-B.

The problem of different models of energy release mentioned in the previous section affects the burnup scale because of the omission of the contributions of fissionable isotopes, mainly Pu-238 and Pu-240. The effect is nearly independent of burnup (see Table 9). The stretching factor of the burnup scale for the results of CEA and EDF is about 1.03 and 1.01 for Benchmark A and B, respectively. Sensitivity calculations of CEA (see Appendix D.1) gave correction values of -392 pcm and -196 pcm for Benchmark A and B respectively, to make the results of CEA and EDF comparable with the others at 50 MWd/kg.

Reactivity change with burnup

Referring to Tables 5-A and 5-B, the reactivity change with burnup in Benchmark A is moderately consistent between the various contributions with a spread of 2.5% Δk at the highest burnup step. When only the results from the codes which are more established in terms of commercial MOX experience are included, the spread reduces to about 1.5% Δk. There is a tendency for those contributions in which Pu-242 self-shielding was not modelled to have the highest reactivity swings (e.g., BEN and BNFL). This may be attributable to the resulting higher levels of Am-243, since Am-243 has a higher absorption cross-section than Pu-242. Table 11 shows that both the BEN and BNFL solutions have the highest absorption rates in Am-243.

For Benchmark B the 2.2% spread in burnup reactivity is only slightly smaller than that of Benchmark A. This implies that the bulk of the discrepancy arises from inherent differences in the depletion characteristics, probably deriving from nuclear library differences and the mutual shielding effect on higher actinide cross-sections.

One-group fluxes

The one-group fluxes also show discrepancies, i.e., spreads of approximately 7% and 4% applying to Benchmarks A and B respectively. This may stem in part from the fact that not all contributors were able to use the specified MeV/fission values, because such a facility is not normally provided in lattice codes. It is surprising that differences exist even for those contributions in which the specified MeV/fission values were used.

Absorption rates

The normalisation of the flux in the cell according to the usual condition "total absorption in the cell equal to unity" ensures that the error in the absorption rate is equivalent to the error in k-infinity, but with opposite sign. Consequently, the macroscopic absorption rates of individual isotopes in the fuel can be used to correlate the differences in k-infinity to individual isotopes. Table 7 lists the actinides with a significant spread in absorption rates (> 1%) between the various contributors at zero irradiation or at 50 MWd/kg. AR denotes the average absorption rate, taking account of all the contributors. dAr denotes the spread of absorption rates about the mean. The largest discrepancies are for Pu-242, consistent with inappropriate treatment of the 2.7 eV resonance in some of the solutions, in which the bulk of the Pu-242 absorptions occurs. Consequently, Benchmark A shows by far the largest discrepancy due to the high absolute concentration of Pu-242.

Relatively large spreads are also noticeable for U-238 in both benchmarks. Since the U-238 cross-sections today can be regarded as well known, it is likely that the resonance absorption calculational methods are responsible for it.

Table 8 shows the corresponding mean absorption rates and spreads for the principal fission products. The absorption rates are for the most part lower than 1%, but the spread of values is often nearly as large as the rates themselves. There is the potential for these spreads to contribute to an uncertainty of up to 1% in k-infinity, and this may arise from a combination of uncertainties in the nuclear cross-section, fission yields and depletion models.

Fission rates and neutrons per fission (ν)

The variations in the normalised fission rate (dFR) have to be multiplied by a factor of about 3 (ν/k-infinity) to obtain the corresponding differences in k-infinity. Table 9 shows the fission rates of the actinides with the highest contribution and the most significant spreads between the various solutions. The largest differences are for Pu-239, Pu-241 and U-238. The reason for the differences seen in U-238 may be due to inadequate cross-section data and to the use of fission spectra, which are not appropriate for the actual fuel composition.

For both benchmarks large variations of ν for minor actinides and differences in the percent range are observed for the main actinides. The spreads on U-238, Pu-239 and Pu-241 are sufficient to cause uncertainties of the order of 0.1% Δk in k-infinity.

Number densities

The discrepancies in number densities are in most cases higher than those for reaction rates, as can be seen from Table 10 which shows percentage differences in number densities of actinides at 50 MWd/kg for both benchmarks. This observation may be explained by the fact that for many isotopes, especially for fission products, a modification of the absorption cross-section causes deviations in the number density of the nuclide with opposite sign yielding only moderate modifications of the absorption rate. The principal actinides fall within a spread of 10 %, except for Pu-242, for which it reaches 25%. The concentrations of Am-243 and the Cm isotopes show similar deviations. The spread for Am-243 must partly be due to the self-shielding issue, as discussed earlier. The minor actinides also show large variations. Overall, the situation is not acceptable, especially for Benchmark A.

Spectra

Participants submitted spectra for Benchmark A and B for five specified burnup steps. It was necessary to re-normalise them to make them comparable. The figures attached to the plots in Appendix C show the spectra both on a double logarithmic scale and on a linear/logarithmic scale. The latter actually show the deviations in flux at the four non-zero burnup steps from the flux calculated at the zero burnup step and clearly show the almost linear evolution of the fast and thermal fluxes with burnup. There are clear differences between the spectra calculated by the various participants. Providing meaningful comment is, however, difficult due to the different group structures used and because some were calculated for the whole cell, while others for the fuel only.

Monte Carlo calculations

ANL and IKE submitted solutions for the fresh fuel state carried out with continuous-energy Monte Carlo codes. The ANL results are given for room temperature only, while the IKE calculations are performed for room temperature and temperatures close to the ones specified in the benchmark (see Appendix B.1 and B.7). The agreement between the solutions for room temperature is not fully satisfactory.

Calculations with APOLLO-2 (see Appendix C.1) indicate, that the combined effect of the temperature discrepancies applied in the MCNP-4 calculations (fuel: -33.2°C, clad: +20.6°C, moderator: -5.9°C) result in too high k-infinity values by 93 pcm and 85 pcm for Benchmark A and B respectively.

Conclusion

The resulting multiplication constants for pin cells of Benchmark A and B show large fluctuations of 3.1% and 1.3% in the fresh state rising to 4.9% and 2.9% at 50 MWd/kg respectively. The solutions with higher dispersion are calculated by commercially established codes, which are mainly applied and verified for uranium fuel. If these solutions are not included, the spread decreases to 0.9% at BOL and 1.5% at 50 MWd/kg. Most participants of this latter group applied new data bases and refined resonance calculations for the generation of shielded resonance cross-sections. It is a similar situation to the one encountered for High Conversion LWR benchmark of OECD/NEA also investigating the behaviour of water-moderated MOX fuel [2,3]. The main resulting recommendations made there are valid for the present benchmark also, and are as follows:

- The calculational methods have to take into account resonance shielding, and should include mutual shielding, over the whole energy region for the fuel and cladding nuclides and the major fission products;
- Sufficient quality of basic nuclear data is needed, in particular for U-238 and the Pu isotopes, but also for higher actinides and fission products.

A part of spread in results of the present benchmark originates from differences of the applied data. Solutions where the new data bases JEF-2, JENDL-3 and ENDF/B-V are used, show characteristic discrepancies for instance in the specific reaction rates, which should be correlated not only to differences in cross-sections of specific isotopes but also to cross-sections in definite energy regions. The energy integrated reaction rates provided in this benchmark do not give sufficient information to allow a detailed evaluation in this respect. Energy-dependent reaction rates would provide guidance to improving the data evaluations and also to refining the methods for calculating weighting spectra and weighted cross-sections.

The large uncertainty related to the minor actinide production is noticed as a by-product of this benchmark. It is caused by differences in the cross-section data bases, but also by insufficient resonance shielding calculations (neglecting the mutual shielding effect). This is clearly shown by the differences of number densities of Pu-242 and its successors Am-243 and Cm-244.

Acknowledgements

Special thanks go to all participants who were willing to devote time to this endeavour and to our colleagues at IKE who provided a great help in processing the results.

References

[1] G. Mangham, "The Generation of Nuclide Macroscopic Data Relating to MOX Recycling Based on the Benchmark Specification Issued by the OECD Working Party on Plutonium Recycling (WPPR)", FEDR 93/2050, BNFL plc, December 1993.

[2] H. Akie, Y. Ishiguro and H. Takano, Summary Report on the "Comparison of NEACRP Burnup Benchmark Calculations for High Conversion Light Water Reactor Lattices", NEACRP-L-309 (1988).

[3] W. Bernnat, Y. Ishiguro, E. Sartori, J. Stepanek and M. Takano, "Advances in the Analysis of the NEACRP High Conversion LWR Benchmark Problems", PHYSOR 90, Marseilles Vol. I, 54 (1990).

Table 1 **Summary of participants**

INSTITUTE	COUNTRY	CODE	DATA BASE/LIBRARY	NO. OF GROUPS	REMARKS
ANL	USA	VIM	ENDF/B-V	infinite	300 K, Zircaloy, no depletion
BEN	Belgium	LWRWIMS	1986 WIMS	69	
BNFL	England	LWRWIMS	1986 WIMS	69	
CEA	France	APOLLO 2	JEF-2.2 CEA 93	172	
ECN	Netherlands	WIMS-D	JEF-2.2 SCALE	172	
EDF	France	APOLLO 1	CEA 86	99	
Hitachi	Japan	VMONT	JENDL-2/ENDF/B-IV	190	
IKE-1	Germany	CGM/RSYST	JEF-1	224/45	
IKE-2	Germany	MCNP 4.2	JEF-2.2	infinite	300 K/600 K, no depletion
JAERI	Japan	SRAC	JENDL-3.1	107	
PSI-1	Switzerland	BOXER	JEF-1	70	
PSI-2	Switzerland	CASMO 3	ENDF/B-IV	40	withdrawn
Siemens	Germany	CASMO 3	J70	70	withdrawn
Studsvik	Sweden	CASMO 4	JEF 2.2	70	

Table 2 **Information about resonance treatment**

ANL: Continuous energy Monte Carlo, shielding of unresolved resonances,

BEN: Self-shielded cross-sections of Pu-239 and Pu-240 (1 eV resonance only),

BNFL: Self-shielded cross-sections of Pu-239 and Pu-240 (1 eV resonance only),

CEA: Self- and mutual shielding for the main U, Pu and Zr isotopes, local and burnup effects included,

ECN: All actinides, fission products and Zr are self shielded at every burnup step,

EDF: Self-shielding and resonance overlapping effect for U-235, U-236, U-238, Pu-239, Pu-240, Pu-241, Pu-242 and Zr isotopes at each burnup step,

HIT: Self-shielding for all actinides and fission products,

IKE 1: Self- and mutual shielding for the main U and Pu isotopes, by performing, an ultrafine group cell calculation, burnup effects included,

IKE 2: Continuous energy Monte Carlo, no special treatment of unresolved resonances,

JAE: Self- and mutual shielding for all U, Pu, Am isotopes and many fission products by performing an ultrafine group cell calculation,

PSI 1: Self- and mutual shielding for U and Pu isotopes by performing an ultrafine group calculation,

PSI 2: All U isotopes and Pu-239 are self shielded,

SIE 1: Self-shielding for the heavy isotopes from U-235 to Am-242m,

STU: Self-shielding for the heavy isotopes from U-235 to Am-242m.

Table 3 **Information about the applied fission yields, energy/fission values and (n,2n)-treatment**

	FISSION YIELDS			ENERGY/FISSION	
	source	nuclide-dep.	5 specif. actinides	all actinides	(n,2n)
ANL					
BEN		yes			
BNFL		yes			FP
CEA	Rider	yes	yes	no	k
ECN		yes	yes	yes	k
EDF		yes	yes	no	k
HIT					
IKE-1	JEF-1	yes	yes	yes	chain
IKE-2					k
JAE	JNDC-90	yes	yes	yes	chain
PSI 1		yes	yes	yes	
PSI 2		yes		yes	k
SIE 1					
STU				yes	k

Table 4-A: k-infinity Benchmark A

Contributor	ANL*	BEN	BNFL	CEA	ECN	EDF	HIT	IKE1	IKE2	JAE	PSI1	STU
Burnup MWd/kg												
0.0	1.1324	1.1044	1.1043	1.1334	1.1313	1.1217	1.1396	1.1308	1.1308	1.1336	1.1304	1.1336
10.0	0.0000	1.0400	1.0398	1.0707	1.0746	1.0593	1.0777	1.0688	0.0000	1.0718	1.0686	1.0747
33.0	0.0000	0.9645	0.9645	0.9974	1.0057	0.9863	1.0081	0.9949	0.0000	1.0028	0.9974	1.0055
42.0	0.0000	0.9405	0.9405	0.9716	0.9821	0.9626	0.9833	0.9705	0.0000	0.9799	0.9743	0.9827
50.0	0.0000	0.9208	0.9208	0.9497	0.9622	0.9433	0.9625	0.9507	0.0000	0.9610	0.9554	0.9641

* The original result of ANL for 300 K is 1.1591 ± 0.0011. It has been converted to required temperatures using the results of IKE2 for room temperature (1.2586 ± 0.0011) and for a set of near benchmark conditions (see Appendix B.7)

Table 4-B: k-infinity Benchmark B

Contributor	ANL*	BNFL	CEA	ECN	EDF	HIT	IKE1	IKE2	JAE	PSI1	STU
Burnup MWd/kg											
0.0	1.1785	1.1805	1.1896	1.1838	1.1744	1.1926	1.1849	1.1847	1.1872	1.1839	1.1830
10.0	0.0000	1.0923	1.0953	1.0972	1.0824	1.1026	1.0936	0.0000	1.0967	1.0929	1.0947
33.0	0.0000	0.9893	0.9876	0.9953	0.9730	0.9956	0.9851	0.0000	0.9931	0.9870	0.9909
42.0	0.0000	0.9541	0.9496	0.9604	0.9367	0.9586	0.9483	0.0000	0.9579	0.9522	0.9560
50.0	0.0000	0.9254	0.9175	0.9312	0.9070	0.9280	0.9178	0.0000	0.9289	0.9241	0.9269

* The original result of ANL for 300 K is 1.2117 ± 00010. It has been converted to required temperatures using the results of IKE2 for room temperature (1.2182 ± 0.0011) and for a set ofnear benchmark conditions (see Appendix B.7)

Table 5-A: Burnup Reactivity Benchmark A

Contributor	BEN	BNFL	CEA	ECN	EDF	HIT	IKE1	JAE	PSI1	STU
Burnup MWd/kg										
0.0	0.0000	0.0000	0.0000	0.0000	0.0000	0.0000	0.0000	0.0000	0.0000	0.0000
10.0	-0.0561	-0.0562	-0.0517	-0.0466	-0.0525	-0.0504	-0.0513	-0.0508	-0.0511	-0.0483
33.0	-0.1313	-0.1313	-0.1203	-0.1104	-0.1224	-0.1145	-0.1208	-0.1151	-0.1180	-0.1124
42.0	-0.1578	-0.1577	-0.1470	-0.1343	-0.1473	-0.1395	-0.1461	-0.1384	-0.1418	-0.1354
50.0	-0.1805	-0.1804	-0.1707	-0.1554	-0.1685	-0.1615	-0.1675	-0.1584	-0.1621	-0.1551

Table 5-B: Burnup Reactivity Benchmark B

Contributor	BNFL	CEA	ECN	EDF	HIT	IKE1	JAE	PSI1	STU
Burnup MWd/kg									
0.0	0.0000	0.0000	0.0000	0.0000	0.0000	0.0000	0.0000	0.0000	0.0000
10.0	-0.0684	-0.0723	-0.0666	-0.0724	-0.0684	-0.0705	-0.0695	-0.0704	-0.0682
33.0	-0.1637	-0.1719	-0.1599	-0.1763	-0.1659	-0.1712	-0.1646	-0.1685	-0.1639
42.0	-0.2011	-0.2125	-0.1964	-0.2162	-0.2047	-0.2106	-0.2016	-0.2056	-0.2007
50.0	-0.2336	-0.2492	-0.2291	-0.2511	-0.2391	-0.2456	-0.2342	-0.2376	-0.2336

Table 6-A: Absolute Fluxes in the Evolution Calculation, $10^{14}/cm^2s$
Benchmark A

Contributor	BEN	BNFL	CEA	ECN	EDF	HIT	IKE1	JAE	PSI1
Burnup MWd/kg									
0.0	3.0701	3.0694	3.1662	2.9375	2.9461	2.9757	2.9017	2.9693	3.0223
10.0	3.2115	3.2115	3.2962	3.0935	3.1218	3.1469	3.0729	3.1373	3.1972
33.0	3.5141	3.5128	3.4254	3.3133	3.3734	3.3826	3.3189	3.3664	3.4406
42.0	3.6075	3.6061	3.5378	3.3947	3.4672	3.4736	3.4120	3.4526	3.5306
50.0	3.7098	3.7081	3.6370	3.4655	3.5491	3.5592	3.4938	3.5282	3.6089

Table 6-B: Absolute Fluxes in the Evolution Calculation, $10^{14}/cm^2s$
Benchmark B

Contributor	BNFL	CEA	ECN	EDF	HIT	IKE1	JAE	PSI1
Burnup MWd/kg								
0.0	2.8468	3.0155	2.8147	2.8185	2.8509	2.7586	2.8466	2.8949
10.0	3.0226	3.1971	3.0418	3.0662	3.0877	2.9961	3.0828	3.1410
33.0	3.4068	3.3901	3.3858	3.4577	3.4642	3.3670	3.4383	3.5162
42.0	3.5410	3.5646	3.5221	3.6147	3.6146	3.5197	3.5822	3.6637
50.0	3.6899	3.7209	3.6436	3.7557	3.7481	3.6563	3.7119	3.7937

Table 7: Actinides with significant uncertainties of absorption rate (AR) in percent, for 0 and 50 MWd/kg

Isotope	Benchmark				Benchmark			
	BOL		50 MWd/kg		BOL		50 MWd/kg	
	AR	dAR	AR	dAR	AR	dAR	AR	dAR
U-235	2.6	0.1	1.5	0.1	1.5	0.05	0.7	0.01
U-238	20.8	0.6	20.3	1.0	23.6	0.7	23.6	1.0
Pu-238	1.2	0.1	1.0	0.2	0.4	0.1	0.4	0.2
Pu-239	36.4	1.2	26.5	0.7	46.1	0.5	27.9	1.2
Pu-240	17.9	2.7	15.0	1.4	13.7	1.0	13.0	0.3
Pu-241	13.4	0.4	14.4	1.3	9.8	0.3	13.7	0.6
Pu-242	5.1	3.6	4.8	2.7	1.2	0.5	2.0	0.6
Am-241	-	-	1.3	0.4	-	-	0.8	0.1
Am-243	-	-	2.9	1.8	-	-	1.3	0.4
Cm-244	-	-	0.6	0.4	-	-	0.2	0.1

Table 8: Fission products with significant uncertainties of absorption rate (AR) in percent at 50 MWd/kg

Isotope	Benchmark A		Benchmark B	
	AR	dAR	AR	dAR
Tc-99	0.5	0.2	0.5	0.2
Rh-103	0.8	0.3	1.2	0.2
Pd-105	0.2	0.1	0.2	0.0
Ag-109	0.4	0.2	0.4	0.3
Xe-131	0.6	0.2	0.7	0.2
Xe-135	0.8	0.1	1.4	0.1
Cs-133	0.6	0.1	0.7	0.2
Nd-143	0.3	0.3	0.6	0.4
Nd-145	0.2	0.1	0.3	0.1
Pm-147	0.4	0.2	0.4	0.2
Sm-149	0.6	0.1	0.7	0.1
Sm-151	0.4	0.1	0.4	0.1
Sm-152	0.4	0.1	0.5	0.2
Eu-153	0.3	0.1	0.4	0.1
Eu-154	0.1	0.1	0.3	0.1
Eu-155	0.2	0.1	0.3	0.1

Table 9: Actinides with significant uncertainties of fission rate (FR) in percent for 0 and 50 MWd/kg

Isotope	Benchmark A				Benchmark B			
	BOL		50 MWd/kg		BOL		50 MWd/kg	
	FR	dFR	FR	dFR	FR	dFR	FR	dFR
U-235	1.9	0.1	1.1	0.06	1.1	0.03	0.6	0.05
U-238	2.8	0.5	2.7	0.5	2.9	0.5	2.8	0.5
Pu-239	23.3	1.0	16.9	0.6	29.6	0.4	17.9	0.7
Pu-240	0.6	0.2	0.5	0.1	0.2	0.06	0.2	0.05
Pu-241	10.1	0.4	10.9	0.5	7.4	0.3	10.3	0.6
Am-242 m	-	-	0.1	0.03	-	-	0.1	0.04
Cm-245	-	-	0.2	0.04	-	-	0.1	0.04

Table 10: Uncertainties in the number densities of actinides in percent at a burnup of 50 MWd/kg

Isotope	Benchmark	
	A	B
U-234	6	67
U-235	3	3
U-236	11	9
U-238	0.1	0.1
Np-237	26	34
Pu-238	14	36
Pu-239	5	8
Pu-240	10	10
Pu-241	11	4
Pu-242	23	22
Am-241	12	19
Am-242m	57	60
A-243	74	33
Cm-242	17	22
Cm-243	26	60
Cm-244	11	14
Cm-245	19	38

Tab. 11: Fract. Absorption Rates of Actinides, Benchmark A

Nuclide	Contributor	Burnup, MWd/kg 0.0	10.0	33.0	42.0	50.0
U-234	BEN	-	-	-	-	-
	BNFL	-	-	-	-	-
	CEA	-	9.086E -05	2.450E -04	2.938E -04	3.343E -04
	ECN	-	1.102E -04	2.835E -04	3.330E -04	3.715E -04
	EDF	-	1.023E -04	2.562E -04	2.955E -04	3.231E -04
	HIT	-	9.343E -05	2.317E -04	2.726E -04	3.021E -04
	IKE1	-	1.015E -04	2.698E -04	3.187E -04	3.569E -04
	IKE2	-	-	-	-	-
	JAE	-	1.045E -04	2.616E -04	3.054E -04	3.393E -04
	PSI1	-	1.052E -04	2.654E -04	3.102E -04	3.447E -04
	STU	-	1.054E -04	2.824E -04	3.339E -04	3.738E -04
	Average	-	1.017E -04	2.619E -04	3.079E -04	3.432E -04
U-235	BEN	2.559E -02	2.269E -02	1.792E -02	1.629E -02	1.492E -02
	BNFL	2.558E -02	2.327E -02	1.832E -02	1.681E -02	1.530E -02
	CEA	2.588E -02	2.286E -02	1.798E -02	1.633E -02	1.495E -02
	ECN	2.596E -02	2.317E -02	1.847E -02	1.685E -02	1.549E -02
	EDF	2.624E -02	2.320E -02	1.815E -02	1.643E -02	1.499E -02
	HIT	2.682E -02	2.379E -02	1.869E -02	1.693E -02	1.547E -02
	IKE1	2.636E -02	2.338E -02	1.842E -02	1.673E -02	1.533E -02
	IKE2	2.600E -02	-	-	-	-
	JAE	2.681E -02	2.371E -02	1.855E -02	1.678E -02	1.531E -02
	PSI1	2.642E -02	2.343E -02	1.854E -02	1.688E -02	1.550E -02
	STU	2.595E -02	2.306E -02	1.830E -02	1.668E -02	1.533E -02
	Average	2.615E -02	2.326E -02	1.834E -02	1.667E -02	1.526E -02
U-236	BEN	-	3.260E -04	9.083E -04	1.084E -03	1.212E -03
	BNFL	-	2.000E -04	8.355E -04	9.992E -04	1.157E -03
	CEA	-	3.418E -04	9.035E -04	1.061E -03	1.178E -03
	ECN	-	3.426E -04	9.108E -04	1.074E -03	1.196E -03
	EDF	-	3.511E -04	9.152E -04	1.071E -03	1.185E -03
	HIT	-	3.412E -04	8.802E -04	1.039E -03	1.147E -03
	IKE1	-	3.807E -04	1.023E -03	1.206E -03	1.348E -03
	IKE2	-	-	-	-	-
	JAE	-	3.914E -04	1.025E -03	1.201E -03	1.331E -03
	PSI1	-	3.561E -04	9.237E -04	1.085E -03	1.207E -03
	STU	-	3.472E -04	9.386E -04	1.109E -03	1.236E -03
	Average	-	3.378E -04	9.264E -04	1.093E -03	1.220E -03

Tab. 11: Fract. Absorption Rates of Actinides, Benchmark A, (cont.)

Nuclide	Contributor	Burnup, MWd/kg 0.0	10.0	33.0	42.0	50.0
U-238	BEN	2.081E -01	2.061E -01	2.041E -01	2.036E -01	2.034E -01
	BNFL	2.081E -01	2.035E -01	2.024E -01	2.015E -01	2.019E -01
	CEA	2.085E -01	2.066E -01	2.042E -01	2.036E -01	2.032E -01
	ECN	2.103E -01	2.093E -01	2.083E -01	2.082E -01	2.081E -01
	EDF	2.061E -01	2.047E -01	2.029E -01	2.024E -01	2.020E -01
	HIT	2.080E -01	2.068E -01	2.046E -01	2.037E -01	2.031E -01
	IKE1	2.087E -01	2.066E -01	2.034E -01	2.024E -01	2.015E -01
	IKE2	2.067E -01	-	-	-	-
	JAE	2.085E -01	2.070E -01	2.052E -01	2.047E -01	2.043E -01
	PSI1	2.100E -01	2.086E -01	2.079E -01	2.081E -01	2.084E -01
	STU	2.045E -01	2.024E -01	1.996E -01	1.988E -01	1.983E -01
	Average	2.080E -01	2.062E -01	2.043E -01	2.037E -01	2.034E -01
Np-237	BEN	-	2.017E -04	7.534E -04	9.814E -04	1.186E -03
	BNFL	-	1.103E -04	6.727E -04	8.705E -04	1.101E -03
	CEA	-	1.840E -04	6.674E -04	8.664E -04	1.044E -03
	ECN	-	1.928E -04	7.124E -04	9.253E -04	1.113E -03
	EDF	-	1.867E -04	6.707E -04	8.677E -04	1.042E -03
	HIT	-	2.108E -04	7.493E -04	9.615E -04	1.150E -03
	IKE1	-	1.649E -04	6.441E -04	8.485E -04	1.033E -03
	IKE2	-	-	-	-	-
	JAE	-	1.580E -04	6.064E -04	7.973E -04	9.681E -04
	PSI1	-	2.149E -04	7.837E -04	1.013E -03	1.215E -03
	STU	-	1.599E -04	5.978E -04	7.836E -04	9.519E -04
	Average	-	1.784E -04	6.858E -04	8.915E -04	1.080E -03
Pu-238	BEN	1.106E -02	1.029E -02	9.345E -03	9.031E -03	8.768E -03
	BNFL	1.106E -02	1.033E -02	9.367E -03	9.064E -03	8.792E -03
	CEA	1.213E -02	1.116E -02	1.035E -02	1.030E -02	1.036E -02
	ECN	1.175E -02	1.095E -02	1.033E -02	1.034E -02	1.043E -02
	EDF	1.188E -02	1.096E -02	1.020E -02	1.016E -02	1.021E -02
	HIT	1.224E -02	1.135E -02	1.055E -02	1.048E -02	1.052E -02
	IKE1	1.175E -02	1.089E -02	1.018E -02	1.015E -02	1.022E -02
	IKE2	1.181E -02	-	-	-	-
	JAE	1.134E -02	1.054E -02	9.980E -03	1.000E -02	1.011E -02
	PSI1	1.182E -02	1.094E -02	1.025E -02	1.021E -02	1.026E -02
	STU	1.177E -02	1.092E -02	1.025E -02	1.025E -02	1.034E -02
	Average	1.169E -02	1.083E -02	1.008E -02	9.997E -03	1.000E -02

Tab. 11: Fract. Absorption Rates of Actinides, Benchmark A, (cont.)

Nuclide	Contributor	Burnup, MWd/kg				
		0.0	10.0	33.0	42.0	50.0
Pu-239	BEN	3.562E -01	3.255E -01	2.860E -01	2.738E -01	2.643E -01
	BNFL	3.562E -01	3.291E -01	2.878E -01	2.759E -01	2.656E -01
	CEA	3.644E -01	3.292E -01	2.858E -01	2.730E -01	2.631E -01
	ECN	3.641E -01	3.332E -01	2.923E -01	2.799E -01	2.703E -01
	EDF	3.682E -01	3.337E -01	2.870E -01	2.728E -01	2.618E -01
	HIT	3.678E -01	3.348E -01	2.916E -01	2.779E -01	2.671E -01
	IKE1	3.656E -01	3.327E -01	2.885E -01	2.747E -01	2.637E -01
	IKE2	3.651E -01	-	-	-	-
	JAE	3.644E -01	3.316E -01	2.885E -01	2.754E -01	2.651E -01
	PSI1	3.632E -01	3.316E -01	2.911E -01	2.792E -01	2.701E -01
	STU	3.660E -01	3.331E -01	2.881E -01	2.741E -01	2.633E -01
	Average	3.638E -01	3.314E -01	2.887E -01	2.757E -01	2.654E -01
Pu-240	BEN	1.611E -01	1.534E -01	1.450E -01	1.428E -01	1.409E -01
	BNFL	1.612E -01	1.533E -01	1.445E -01	1.421E -01	1.404E -01
	CEA	1.823E -01	1.724E -01	1.577E -01	1.530E -01	1.490E -01
	ECN	1.823E -01	1.739E -01	1.617E -01	1.577E -01	1.541E -01
	EDF	1.847E -01	1.750E -01	1.604E -01	1.555E -01	1.513E -01
	HIT	1.799E -01	1.716E -01	1.593E -01	1.555E -01	1.519E -01
	IKE1	1.800E -01	1.714E -01	1.585E -01	1.541E -01	1.501E -01
	IKE2	1.839E -01	-	-	-	-
	JAE	1.842E -01	1.751E -01	1.614E -01	1.567E -01	1.525E -01
	PSI1	1.799E -01	1.713E -01	1.592E -01	1.553E -01	1.520E -01
	STU	1.879E -01	1.784E -01	1.639E -01	1.589E -01	1.546E -01
	Average	1.789E -01	1.696E -01	1.572E -01	1.532E -01	1.497E -01
Pu-241	BEN	1.334E -01	1.327E -01	1.358E -01	1.365E -01	1.370E -01
	BNFL	1.334E -01	1.299E -01	1.343E -01	1.348E -01	1.358E -01
	CEA	1.334E -01	1.230E -01	1.449E -01	1.453E -01	1.448E -01
	ECN	1.333E -01	1.368E -01	1.452E -01	1.468E -01	1.474E -01
	EDF	1.299E -01	1.335E -01	1.421E -01	1.434E -01	1.437E -01
	HIT	1.313E -01	1.343E -01	1.429E -01	1.442E -01	1.446E -01
	IKE1	1.338E -01	1.363E -01	1.429E -01	1.438E -01	1.441E -01
	IKE2	1.331E -01	-	-	-	-
	JAE	1.340E -01	1.371E -01	1.449E -01	1.460E -01	1.462E -01
	PSI1	1.341E -01	1.361E -01	1.430E -01	1.442E -01	1.448E -01
	STU	1.335E -01	1.373E -01	1.463E -01	1.479E -01	1.484E -01
	Average	1.330E -01	1.337E -01	1.422E -01	1.433E -01	1.437E -01

Tab. 11: Fract. Absorption Rates of Actinides, Benchmark A, (cont.)

Nuclide	Contributor	Burnup, MWd/kg				
		0.0	10.0	33.0	42.0	50.0
Pu-242	BEN	7.752E -02	7.490E -02	6.999E -02	6.829E -02	6.681E -02
	BNFL	7.759E -02	7.472E -02	6.999E -02	6.832E -02	6.687E -02
	CEA	4.556E -02	4.470E -02	4.364E -02	4.341E -02	4.324E -02
	ECN	4.570E -02	4.502E -02	4.427E -02	4.415E -02	4.407E -02
	EDF	4.787E -02	4.685E -02	4.545E -02	4.508E -02	4.481E -02
	HIT	4.678E -02	4.585E -02	4.430E -02	4.426E -02	4.415E -02
	IKE1	4.502E -02	4.422E -02	4.317E -02	4.307E -02	4.286E -02
	IKE2	4.526E -02	-	-	-	-
	JAE	4.461E -02	4.393E -02	4.338E -02	4.337E -02	4.341E -02
	PSI1	4.506E -02	4.433E -02	4.366E -02	4.360E -02	4.360E -02
	STU	4.165E -02	4.103E -02	4.035E -02	4.025E -02	4.020E -02
	Average	5.115E -02	5.056E -02	4.882E -02	4.838E -02	4.800E -02
Am-241	BEN	-	3.811E -03	9.720E -03	1.120E -02	1.217E -02
	BNFL	-	2.359E -03	9.057E -03	1.048E -02	1.176E -02
	CEA	-	3.539E -03	9.911E -03	1.178E -02	1.316E -02
	ECN	-	4.209E -03	1.089E -02	1.258E -02	1.372E -02
	EDF	-	4.171E -03	1.080E -02	1.248E -02	1.359E -02
	HIT	-	3.862E -03	1.020E -02	1.186E -02	1.302E -02
	IKE1	-	4.153E -03	1.067E -02	1.229E -02	1.338E -02
	IKE2	-	-	-	-	-
	JAE	-	3.941E -03	1.037E -02	1.204E -02	1.318E -02
	PSI1	-	3.811E -03	9.977E -03	1.158E -02	1.268E -02
	STU	-	4.233E -03	1.106E -02	1.279E -02	1.394E -02
	Average	-	3.809E -03	1.027E -02	1.191E -02	1.306E -02
Am-242m	BEN	-	1.590E -04	8.994E -04	1.151E -03	1.330E -03
	BNFL	-	6.380E -05	7.988E -04	1.034E -03	1.259E -03
	CEA	-	1.477E -04	8.685E -04	1.135E -03	1.341E -03
	ECN	-	1.513E -04	8.399E -04	1.071E -03	1.237E -03
	EDF	-	1.624E -04	9.322E -04	1.198E -03	1.390E -03
	HIT	-	-	-	-	-
	IKE1	-	1.468E -04	8.206E -04	1.046E -03	1.210E -03
	IKE2	-	-	-	-	-
	JAE	-	2.120E -04	1.239E -03	1.601E -03	1.867E -03
	PSI1	-	-	-	-	-
	STU	-	1.653E -04	9.341E -04	1.194E -03	1.383E -03
	Average	-	1.510E -04	9.165E -04	1.179E -03	1.377E -03

Tab. 11: Fract. Absorption Rates of Actinides, Benchmark A, (cont.)

Nuclide	Contributor	Burnup, MWd/kg				
		0.0	10.0	33.0	42.0	50.0
Am-243	BEN	-	1.445E -02	3.437E -02	3.953E -02	4.332E -02
	BNFL	-	9.207E -03	3.212E -02	3.698E -02	4.167E -02
	CEA	-	8.747E -03	2.093E -02	2.413E -02	2.653E -02
	ECN	-	8.563E -03	2.084E -02	2.415E -02	2.665E -02
	EDF	-	9.021E -03	2.162E -02	2.491E -02	2.736E -02
	HIT	-	8.259E -03	2.022E -02	2.338E -02	2.592E -02
	IKE1	-	8.208E -03	2.003E -02	2.319E -02	2.567E -02
	IKE2	-	-	-	-	-
	JAE	-	8.006E -03	1.953E -02	2.270E -02	2.513E -02
	PSI1	-	7.790E -03	1.942E -02	2.271E -02	2.524E -02
	STU	-	7.957E -03	1.956E -02	2.265E -02	2.498E -02
	Average	-	9.021E -03	2.286E -02	2.643E -02	2.925E -02
Cm-242	BEN	-	-	-	-	-
	BNFL	-	-	-	-	-
	CEA	-	3.454E -05	1.900E -04	2.418E -04	2.796E -04
	ECN	-	3.415E -05	1.793E -04	2.295E -04	2.677E -04
	EDF	-	4.183E -05	2.192E -04	2.813E -04	3.296E -04
	HIT	-	3.161E -05	1.649E -04	2.121E -04	2.484E -04
	IKE1	-	3.367E -05	1.758E -04	2.248E -04	2.625E -04
	IKE2	-	-	-	-	-
	JAE	-	3.641E -05	1.941E -04	2.496E -04	2.923E -04
	PSI1	-	3.102E -05	1.630E -04	2.096E -04	2.440E -04
	STU	-	3.477E -05	1.849E -04	2.374E -04	2.779E -04
	Average	-	3.475E -05	1.839E -04	2.358E -04	2.752E -04
Cm-243	BEN	-	-	-	-	-
	BNFL	-	-	-	-	-
	CEA	-	3.875E -06	6.568E -05	1.027E -04	1.364E -04
	ECN	-	3.060E -06	5.227E -05	8.311E -05	1.125E -04
	EDF	-	3.810E -06	6.374E -05	1.011E -04	1.367E -04
	HIT	-	-	-	-	-
	IKE1	-	3.197E -06	5.297E -05	8.373E -05	1.130E -04
	IKE2	-	-	-	-	-
	JAE	-	2.575E -06	4.625E -05	7.464E -05	1.023E -04
	PSI1	-	2.968E -06	4.898E -05	7.779E -05	1.051E -04
	STU	-	3.096E -06	5.305E -05	8.453E -05	1.148E -04
	Average	-	3.226E -06	5.470E -05	8.680E -05	1.173E -04

Tab. 11: Fract. Absorption Rates of Actinides, Benchmark A, (cont.)

Nuclide	Contributor	Burnup, MWd/kg				
		0.0	10.0	33.0	42.0	50.0
Cm-244	BEN	-	-	-	-	-
	BNFL	-	-	-	-	-
	CEA	-	4.601E -04	3.519E -03	5.038E -03	6.437E -03
	ECN	-	-	-	-	-
	EDF	-	-	-	-	-
	HIT	-	3.617E -04	2.660E -03	3.835E -03	4.897E -03
	IKE1	-	3.525E -04	2.834E -03	4.146E -03	5.424E -03
	IKE2	-	-	-	-	-
	JAE	-	4.052E -04	3.093E -03	4.408E -03	5.607E -03
	PSI1	-	-	-	-	-
	STU	-	4.022E -04	3.249E -03	4.748E -03	6.182E -03
	Average	-	3.963E -04	3.071E -03	4.435E -03	5.709E -03
Cm-245	BEN	-	-	-	-	-
	BNFL	-	-	-	-	-
	CEA	-	-	-	-	-
	ECN	-	-	-	-	-
	EDF	-	-	-	-	-
	HIT	-	2.508E -05	6.630E -04	1.198E -03	1.806E -03
	IKE1	-	2.714E -05	7.119E -04	1.315E -03	2.023E -03
	IKE2	-	-	-	-	-
	JAE	-	2.943E -05	7.650E -04	1.389E -03	2.095E -03
	PSI1	-	-	-	-	-
	STU	-	3.051E -05	8.136E -04	1.504E -03	2.316E -03
	Average	-	2.804E -05	7.384E -04	1.352E -03	2.060E -03

Tab. 12: Absorption Rates of Fission Products, Benchmark A

Nuclide	Contributor	Burnup, MWd/kg				
		0.0	10.0	33.0	42.0	50.0
Mo-95	BEN	-	3.426E -04	1.103E -03	1.394E -03	1.650E -03
	BNFL	-	2.038E -04	9.962E -04	1.251E -03	1.542E -03
	CEA	-	1.808E -04	1.094E -03	1.445E -03	1.750E -03
	ECN	-	2.110E -04	1.087E -03	1.407E -03	1.679E -03
	EDF	-	-	-	-	-
	HIT	-	2.126E -04	1.051E -03	1.360E -03	1.632E -03
	IKE1	-	1.974E -04	1.073E -03	1.412E -03	1.706E -03
	JAE	-	4.012E -04	1.158E -03	1.416E -03	1.633E -03
	PSI1	-	1.989E -04	1.092E -03	1.436E -03	1.734E -03
	STU	-	-	-	-	-
	Average	-	2.435E -04	1.082E -03	1.390E -03	1.666E -03
Tc-99	BEN	-	6.559E -04	2.062E -03	2.587E -03	3.044E -03
	BNFL	-	3.924E -04	1.867E -03	2.329E -03	2.850E -03
	CEA	-	1.032E -03	3.077E -03	3.782E -03	4.376E -03
	ECN	-	8.728E -04	2.572E -03	3.153E -03	3.639E -03
	EDF	-	-	-	-	-
	HIT	-	1.030E -03	3.029E -03	3.735E -03	4.304E -03
	IKE1	-	1.030E -03	3.155E -03	3.905E -03	4.543E -03
	JAE	-	1.022E -03	2.999E -03	3.672E -03	4.237E -03
	PSI1	-	1.110E -03	3.245E -03	3.975E -03	4.590E -03
	STU	-	-	-	-	-
	Average	-	8.931E -04	2.751E -03	3.392E -03	3.948E -03
Ru-101	BEN	-	2.434E -04	7.977E -04	1.015E -03	1.209E -03
	BNFL	-	-	-	-	-
	CEA	-	5.134E -04	1.652E -03	2.082E -03	2.458E -03
	ECN	-	-	-	-	-
	EDF	-	-	-	-	-
	HIT	-	4.643E -04	1.477E -03	1.856E -03	2.189E -03
	IKE1	-	4.476E -04	1.448E -03	1.829E -03	2.164E -03
	JAE	-	4.625E -04	1.473E -03	1.850E -03	2.177E -03
	PSI1	-	-	-	-	-
	STU	-	-	-	-	-
	Average	-	4.262E -04	1.370E -03	1.726E -03	2.039E -03

Tab. 12: Absorption Rates of Fission Products, Benchmark A, (cont.)

Nuclide	Contributor	Burnup, MWd/kg				
		0.0	10.0	33.0	42.0	50.0
Rh-103	BEN	-	1.036E -03	3.546E -03	4.440E -03	5.218E -03
	BNFL	-	-	-	-	-
	CEA	-	1.702E -03	5.884E -03	7.254E -03	8.388E -03
	ECN	-	-	-	-	-
	EDF	-	-	-	-	-
	HIT	-	1.682E -03	5.599E -03	6.918E -03	8.011E -03
	IKE1	-	1.646E -03	5.463E -03	6.713E -03	7.766E -03
	JAE	-	1.683E -03	5.633E -03	6.941E -03	8.028E -03
	PSI1	-	-	-	-	-
	STU	-	2.214E -03	6.312E -03	7.689E -03	8.832E -03
	Average	-	1.660E -03	5.406E -03	6.659E -03	7.707E -03
Pd-105	BEN	-	2.128E -04	7.035E -04	8.972E -04	1.071E -03
	BNFL	-	-	-	-	-
	CEA	-	4.312E -04	1.422E -03	1.805E -03	2.143E -03
	ECN	-	-	-	-	-
	EDF	-	-	-	-	-
	HIT	-	4.186E -04	1.363E -03	1.727E -03	2.044E -03
	IKE1	-	4.045E -04	1.329E -03	1.686E -03	2.002E -03
	JAE	-	4.120E -04	1.346E -03	1.704E -03	2.019E -03
	PSI1	-	-	-	-	-
	STU	-	-	-	-	-
	Average	-	3.758E -04	1.233E -03	1.564E -03	1.856E -03
Pd-107	BEN	-	-	-	-	-
	BNFL	-	-	-	-	-
	CEA	-	3.136E -04	1.023E -03	1.294E -03	1.531E -03
	ECN	-	-	-	-	-
	EDF	-	-	-	-	-
	HIT	-	3.007E -04	9.662E -04	1.221E -03	1.445E -03
	IKE1	-	2.995E -04	9.849E -04	1.252E -03	1.490E -03
	JAE	-	3.173E -04	1.025E -03	1.293E -03	1.526E -03
	PSI1	-	-	-	-	-
	STU	-	-	-	-	-
	Average	-	3.078E -04	9.998E -04	1.265E -03	1.498E -03

Tab. 12: Absorption Rates of Fission Products, Benchmark A, (cont.)

Nuclide	Contributor	Burnup, MWd/kg				
		0.0	10.0	33.0	42.0	50.0
Pd-108	BEN	-	3.425E -04	1.083E -03	1.357E -03	1.593E -03
	BNFL	-	-	-	-	-
	CEA	-	3.192E -04	1.065E -03	1.359E -03	1.621E -03
	ECN	-	-	-	-	-
	EDF	-	-	-	-	-
	HIT	-	3.613E -04	1.102E -03	1.382E -03	1.624E -03
	IKE1	-	3.279E -04	1.093E -03	1.396E -03	1.669E -03
	JAE	-	4.097E -04	1.190E -03	1.459E -03	1.687E -03
	PSI1	-	-	-	-	-
	STU	-	-	-	-	-
	Average	-	3.521E -04	1.107E -03	1.390E -03	1.639E -03
Ag-109	BEN	-	8.970E -04	2.581E -03	3.140E -03	3.602E -03
	BNFL	-	5.462E -04	2.362E -03	2.866E -03	3.405E -03
	CEA	-	1.297E -03	3.352E -03	3.961E -03	4.445E -03
	ECN	-	6.690E -04	1.688E -03	1.973E -03	2.192E -03
	EDF	-	-	-	-	-
	HIT	-	1.117E -03	2.916E -03	3.500E -03	3.934E -03
	IKE1	-	1.305E -03	3.532E -03	4.206E -03	4.742E -03
	JAE	-	1.375E -03	3.529E -03	4.153E -03	4.642E -03
	PSI1	-	1.286E -03	3.361E -03	4.008E -03	4.537E -03
	STU	-	1.231E -03	3.136E -03	3.636E -03	3.990E -03
	Average	-	1.080E -03	2.940E -03	3.494E -03	3.943E -03
Xe-131	BEN	-	1.480E -03	4.225E -03	5.091E -03	5.772E -03
	BNFL	-	8.993E -04	3.876E -03	4.667E -03	5.481E -03
	CEA	-	1.752E -03	4.972E -03	5.915E -03	6.635E -03
	ECN	-	1.406E -03	3.792E -03	4.481E -03	5.005E -03
	EDF	-	-	-	-	-
	HIT	-	1.433E -03	3.979E -03	4.747E -03	5.380E -03
	IKE1	-	1.765E -03	5.012E -03	5.957E -03	6.677E -03
	JAE	-	1.483E -03	3.579E -03	4.177E -03	4.647E -03
	PSI1	-	1.594E -03	4.370E -03	5.221E -03	5.896E -03
	STU	-	1.866E -03	5.163E -03	6.140E -03	6.877E -03
	Average	-	1.520E -03	4.330E -03	5.155E -03	5.819E -03

Tab. 12: Absorption Rates of Fission Products, Benchmark A, (cont.)

Nuclide	Contributor	Burnup, MWd/kg				
		0.0	10.0	33.0	42.0	50.0
Xe-135	BEN	-	6.647E -03	7.077E -03	7.264E -03	7.494E -03
	BNFL	-	-	-	-	-
	CEA	-	8.340E -03	7.908E -03	7.814E -03	7.752E -03
	ECN	-	-	-	-	-
	EDF	-	-	-	-	-
	HIT	-	7.445E -03	7.817E -03	8.005E -03	8.167E -03
	IKE1	-	7.469E -03	7.805E -03	7.963E -03	8.115E -03
	JAE	-	7.486E -03	7.832E -03	8.005E -03	8.166E -03
	PSI1	-	-	-	-	-
	STU	-	7.238E -03	7.602E -03	7.762E -03	7.960E -03
	Average	-	7.438E -03	7.673E -03	7.802E -03	7.942E -03
Cs-133	BEN	-	1.341E -03	4.078E -03	5.049E -03	5.872E -03
	BNFL	-	8.072E -04	3.708E -03	4.573E -03	5.523E -03
	CEA	-	1.620E -03	4.754E -03	5.787E -03	6.638E -03
	ECN	-	1.382E -03	4.064E -03	4.960E -03	5.682E -03
	EDF	-	-	-	-	-
	HIT	-	1.364E -03	4.093E -03	5.018E -03	5.765E -03
	IKE1	-	1.597E -03	4.816E -03	5.908E -03	6.817E -03
	JAE	-	1.380E -03	3.935E -03	4.771E -03	5.460E -03
	PSI1	-	1.467E -03	4.372E -03	5.353E -03	6.171E -03
	STU	-	1.688E -03	4.977E -03	6.088E -03	7.000E -03
	Average	-	1.405E -03	4.311E -03	5.278E -03	6.103E -03
Cs-135	BEN	-	2.550E -04	8.143E -04	1.023E -03	1.204E -03
	BNFL	-	-	-	-	-
	CEA	-	2.709E -04	8.880E -04	1.125E -03	1.333E -03
	ECN	-	-	-	-	-
	EDF	-	-	-	-	-
	HIT	-	2.324E -04	7.246E -04	9.059E -04	1.065E -03
	IKE1	-	2.861E -04	9.234E -04	1.163E -03	1.370E -03
	JAE	-	2.630E -04	8.090E -04	1.002E -03	1.167E -03
	PSI1	-	-	-	-	-
	STU	-	2.690E -04	8.661E -04	1.089E -03	1.282E -03
	Average	-	2.627E -04	8.376E -04	1.051E -03	1.237E -03

Tab. 12: Absorption Rates of Fission Products, Benchmark A, (cont.)

Nuclide	Contributor	Burnup, MWd/kg				
		0.0	10.0	33.0	42.0	50.0
Nd-143	BEN	-	4.556E -04	1.543E -03	2.000E -03	2.425E -03
	BNFL	-	2.697E -04	1.381E -03	1.773E -03	2.244E -03
	CEA	-	6.346E -04	2.229E -03	2.859E -03	3.425E -03
	ECN	-	6.883E -04	2.416E -03	3.104E -03	3.721E -03
	EDF	-	-	-	-	-
	HIT	-	8.897E -04	3.167E -03	4.100E -03	4.935E -03
	IKE1	-	6.095E -04	2.131E -03	2.731E -03	3.269E -03
	JAE	-	6.258E -04	2.142E -03	2.732E -03	3.257E -03
	PSI1	-	6.069E -04	2.103E -03	2.694E -03	3.224E -03
	STU	-	6.701E -04	2.200E -03	2.806E -03	3.346E -03
	Average	-	6.056E -04	2.146E -03	2.755E -03	3.316E -03
Nd-145	BEN	-	-	-	-	-
	BNFL	-	2.580E -04	1.224E -03	1.525E -03	1.863E -03
	CEA	-	5.055E -04	1.573E -03	1.963E -03	2.298E -03
	ECN	-	5.841E -04	1.850E -03	2.321E -03	2.729E -03
	EDF	-	-	-	-	-
	HIT	-	4.299E -04	1.319E -03	1.645E -03	1.930E -03
	IKE1	-	4.941E -04	1.550E -03	1.937E -03	2.268E -03
	JAE	-	4.399E -04	1.333E -03	1.650E -03	1.921E -03
	PSI1	-	4.617E -04	1.436E -03	1.795E -03	2.107E -03
	STU	-	4.886E -04	1.513E -03	1.884E -03	2.200E -03
	Average	-	4.577E -04	1.475E -03	1.840E -03	2.165E -03
Pm-147	BEN	-	1.066E -03	2.004E -03	2.121E -03	2.175E -03
	BNFL	-	7.065E -04	1.927E -03	2.055E -03	2.141E -03
	CEA	-	1.901E -03	3.798E -03	4.023E -03	4.108E -03
	ECN	-	2.009E -03	4.043E -03	4.356E -03	4.530E -03
	EDF	-	-	-	-	-
	HIT	-	1.768E -03	3.545E -03	3.847E -03	3.985E -03
	IKE1	-	1.917E -03	3.920E -03	4.204E -03	4.349E -03
	JAE	-	1.755E -03	3.490E -03	3.767E -03	3.924E -03
	PSI1	-	1.848E -03	3.663E -03	3.945E -03	4.104E -03
	STU	-	2.090E -03	4.060E -03	4.331E -03	4.454E -03
	Average	-	1.673E -03	3.383E -03	3.628E -03	3.752E -03

Tab. 12: Absorption Rates of Fission Products, Benchmark A, (cont.)

Nuclide	Contributor	Burnup, MWd/kg				
		0.0	10.0	33.0	42.0	50.0
Pm-148m	BEN	-	6.110E -04	1.271E -03	1.370E -03	1.425E -03
	BNFL	-	3.742E -04	1.213E -03	1.317E -03	1.396E -03
	CEA	-	3.384E -04	7.736E -04	8.324E -04	8.608E -04
	ECN	-	3.040E -04	7.372E -04	8.311E -04	8.898E -04
	EDF	-	-	-	-	-
	HIT	-	2.945E -04	7.004E -04	7.825E -04	8.408E -04
	IKE1	-	3.094E -04	7.742E -04	8.670E -04	9.318E -04
	JAE	-	3.159E -04	7.570E -04	8.560E -04	9.207E -04
	PSI1	-	3.097E -04	7.668E -04	8.493E -04	9.055E -04
	STU	-	3.421E -04	8.120E -04	9.023E -04	9.690E -04
	Average	-	3.555E -04	8.672E -04	9.564E -04	1.016E -03
Sm-149	BEN	-	5.493E -03	6.316E -03	6.384E -03	6.371E -03
	BNFL	-	4.794E -03	6.233E -03	6.311E -03	6.353E -03
	CEA	-	5.310E -03	5.766E -03	5.787E -03	5.778E -03
	ECN	-	4.626E -03	5.078E -03	5.166E -03	5.210E -03
	EDF	-	-	-	-	-
	HIT	-	5.490E -03	5.910E -03	5.983E -03	6.016E -03
	IKE1	-	5.373E -03	5.849E -03	5.913E -03	5.957E -03
	JAE	-	5.350E -03	5.772E -03	5.861E -03	5.906E -03
	PSI1	-	5.364E -03	5.878E -03	5.975E -03	6.031E -03
	STU	-	5.385E -03	5.773E -03	5.771E -03	5.805E -03
	Average	-	5.243E -03	5.842E -03	5.906E -03	5.936E -03
Sm-150	BEN	-	1.878E -04	8.421E -04	1.113E -03	1.356E -03
	BNFL	-	8.857E -05	7.454E -04	9.814E -04	1.255E -03
	CEA	-	2.395E -04	9.583E -04	1.242E -03	1.491E -03
	ECN	-	2.002E -04	8.039E -04	1.046E -03	1.261E -03
	EDF	-	-	-	-	-
	HIT	-	1.958E -04	7.391E -04	9.524E -04	1.150E -03
	IKE1	-	2.289E -04	9.184E -04	1.193E -03	1.439E -03
	JAE	-	2.300E -04	8.705E -04	1.110E -03	1.317E -03
	PSI1	-	1.174E -04	4.933E -04	6.534E -04	8.009E -04
	STU	-	2.754E -04	1.079E -03	1.389E -03	1.657E -03
	Average	-	1.960E -04	8.278E -04	1.076E -03	1.303E -03

Tab. 12: Absorption Rates of Fission Products, Benchmark A, (cont.)

Nuclide	Contributor	Burnup, MWd/kg				
		0.0	10.0	33.0	42.0	50.0
Sm-151	BEN	-	1.113E -03	2.739E -03	3.194E -03	3.549E -03
	BNFL	-	7.034E -04	2.547E -03	2.967E -03	3.394E -03
	CEA	-	1.851E -03	3.456E -03	3.788E -03	4.046E -03
	ECN	-	1.336E -03	2.513E -03	2.769E -03	2.971E -03
	EDF	-	-	-	-	-
	HIT	-	1.871E -03	3.375E -03	3.632E -03	3.810E -03
	IKE1	-	1.862E -03	3.470E -03	3.791E -03	4.043E -03
	JAE	-	1.852E -03	3.424E -03	3.718E -03	3.933E -03
	PSI1	-	1.749E -03	3.029E -03	3.223E -03	3.367E -03
	STU	-	1.823E -03	3.470E -03	3.820E -03	4.093E -03
	Average	-	1.573E -03	3.114E -03	3.433E -03	3.690E -03
Sm-152	BEN	-	8.641E -04	2.782E -03	3.431E -03	3.956E -03
	BNFL	-	5.055E -04	2.526E -03	3.115E -03	3.734E -03
	CEA	-	1.219E -03	3.631E -03	4.253E -03	4.698E -03
	ECN	-	8.439E -04	2.540E -03	2.990E -03	3.319E -03
	EDF	-	-	-	-	-
	HIT	-	9.940E -04	3.003E -03	3.611E -03	4.027E -03
	IKE1	-	1.024E -03	3.311E -03	3.974E -03	4.484E -03
	JAE	-	9.047E -04	2.773E -03	3.341E -03	3.782E -03
	PSI1	-	1.242E -03	3.535E -03	4.108E -03	4.507E -03
	STU	-	1.207E -03	3.690E -03	4.384E -03	4.883E -03
	Average	-	9.783E -04	3.088E -03	3.690E -03	4.154E -03
Eu-153	BEN	-	3.970E -04	1.700E -03	2.309E -03	2.872E -03
	BNFL	-	2.211E -04	1.492E -03	2.012E -03	2.638E -03
	CEA	-	4.742E -04	2.143E -03	2.867E -03	3.493E -03
	ECN	-	3.026E -04	1.396E -03	1.880E -03	2.302E -03
	EDF	-	-	-	-	-
	HIT	-	4.347E -04	1.858E -03	2.485E -03	3.041E -03
	IKE1	-	4.298E -04	1.907E -03	2.573E -03	3.169E -03
	JAE	-	4.206E -04	1.758E -03	2.344E -03	2.863E -03
	PSI1	-	4.760E -04	2.159E -03	2.875E -03	3.479E -03
	STU	-	4.883E -04	2.169E -03	2.911E -03	3.561E -03
	Average	-	4.049E -04	1.842E -03	2.473E -03	3.047E -03

Tab. 12: Absorption Rates of Fission Products, Benchmark A, (cont.)

Nuclide	Contributor	Burnup, MWd/kg				
		0.0	10.0	33.0	42.0	50.0
Eu-154	BEN	-	2.679E -05	3.566E -04	6.147E -04	9.127E -04
	BNFL	-	9.142E -06	2.851E -04	4.845E -04	7.866E -04
	CEA	-	6.808E -05	7.931E -04	1.266E -03	1.746E -03
	ECN	-	4.205E -05	5.012E -04	8.080E -04	1.123E -03
	EDF	-	-	-	-	-
	HIT	-	5.209E -05	7.694E -04	1.221E -03	1.685E -03
	IKE1	-	6.029E -05	6.798E -04	1.092E -03	1.521E -03
	JAE	-	6.390E -05	6.948E -04	1.104E -03	1.524E -03
	PSI1	-	6.320E -05	7.585E -04	1.216E -03	1.679E -03
	STU	-	7.012E -05	7.926E -04	1.267E -03	1.754E -03
	Average	-	5.063E -05	6.257E -04	1.008E -03	1.415E -03
Eu-155	BEN	-	5.050E -04	8.723E -04	1.081E -03	1.336E -03
	BNFL	-	3.701E -04	8.110E -04	9.676E -04	1.222E -03
	CEA	-	4.047E -04	1.051E -03	1.423E -03	1.827E -03
	ECN	-	2.338E -04	6.084E -04	8.379E -04	1.095E -03
	EDF	-	-	-	-	-
	HIT	-	6.372E -04	1.224E -03	1.621E -03	2.063E -03
	IKE1	-	3.708E -04	9.412E -04	1.260E -03	1.614E -03
	JAE	-	6.185E -04	1.158E -03	1.515E -03	1.898E -03
	PSI1	-	4.903E -04	1.183E -03	1.549E -03	1.945E -03
	STU	-	5.160E -04	1.225E -03	1.594E -03	2.000E -03
	Average	-	4.607E -04	1.008E -03	1.316E -03	1.667E -03

Tab. 13: Fractional Fission Rate of Actinides, Benchmark A

Nuclide	Contributor	Burnup, MWd/kg				
		0.0	10.0	33.0	42.0	50.0
U-234	BEN	-	-	-	-	-
	BNFL	-	-	-	-	-
	CEA	-	3.353E -06	9.842E -06	1.203E -05	1.387E -05
	ECN	-	3.892E -06	1.061E -05	1.261E -05	1.417E -05
	EDF	-	3.809E -06	1.029E -05	1.210E -05	1.341E -05
	HIT	-	3.784E -06	1.047E -05	1.250E -05	1.411E -05
	IKE1	-	3.817E -06	1.048E -05	1.247E -05	1.401E -05
	IKE2	-	-	-	-	-
	JAE	-	3.999E -06	1.098E -05	1.310E -05	1.479E -05
	PSI1	-	3.691E -06	1.005E -05	1.196E -05	1.347E -05
	STU	-	4.487E -06	1.242E -05	1.478E -05	1.663E -05
	Average	-	3.854E -06	1.064E -05	1.270E -05	1.431E -05
U-235	BEN	1.867E -02	1.653E -02	1.313E -02	1.197E -02	1.099E -02
	BNFL	1.866E -02	1.694E -02	1.341E -02	1.234E -02	1.126E -02
	CEA	1.925E -02	1.698E -02	1.417E -02	1.223E -02	1.122E -02
	ECN	1.930E -02	1.721E -02	1.379E -02	1.260E -02	1.162E -02
	EDF	1.931E -02	1.705E -02	1.341E -02	1.216E -02	1.112E -02
	HIT	1.972E -02	1.747E -02	1.379E -02	1.253E -02	1.147E -02
	IKE1	1.937E -02	1.717E -02	1.360E -02	1.238E -02	1.138E -02
	IKE2	1.938E -02	-	-	-	-
	JAE	1.924E -02	1.700E -02	1.338E -02	1.213E -02	1.109E -02
	PSI1	1.946E -02	1.723E -02	1.370E -02	1.250E -02	1.150E -02
	STU	1.926E -02	1.711E -02	1.366E -02	1.248E -02	1.150E -02
	Average	1.924E -02	1.707E -02	1.360E -02	1.233E -02	1.132E -02
U-236	BEN	-	1.073E -05	3.082E -05	3.699E -05	4.174E -05
	BNFL	-	6.492E -06	2.830E -05	3.399E -05	3.973E -05
	CEA	-	1.480E -05	4.126E -05	4.918E -05	5.521E -05
	ECN	-	1.451E -05	4.066E -05	4.859E -05	5.467E -05
	EDF	-	1.459E -05	4.059E -05	4.835E -05	5.424E -05
	HIT	-	1.479E -05	4.115E -05	4.909E -05	5.496E -05
	IKE1	-	1.514E -05	4.234E -05	5.044E -05	5.664E -05
	IKE2	-	-	-	-	-
	JAE	-	1.674E -05	4.626E -05	5.492E -05	6.144E -05
	PSI1	-	1.478E -05	4.110E -05	4.912E -05	5.529E -05
	STU	-	1.664E -05	4.682E -05	5.593E -05	6.289E -05
	Average	-	1.392E -05	3.993E -05	4.766E -05	5.368E -05

Tab. 13: Fractional Fission Rate of Actinides, Benchmark A, (cont.)

Nuclide	Contributor	Burnup, MWd/kg				
		0.0	10.0	33.0	42.0	50.0
U-238	BEN	2.762E -02	2.753E -02	2.728E -02	2.718E -02	2.709E -02
	BNFL	2.764E -02	2.714E -02	2.709E -02	2.694E -02	2.693E -02
	CEA	2.778E -02	2.763E -02	2.730E -02	2.717E -02	2.706E -02
	ECN	2.706E -02	2.691E -02	2.661E -02	2.651E -02	2.639E -02
	EDF	2.637E -02	2.625E -02	2.597E -02	2.587E -02	2.577E -02
	HIT	2.711E -02	2.688E -02	2.659E -02	2.637E -02	2.618E -02
	IKE1	2.725E -02	2.715E -02	2.686E -02	2.675E -02	2.665E -02
	IKE2	2.705E -02	-	-	-	-
	JAE	2.839E -02	2.805E -02	2.753E -02	2.733E -02	2.715E -02
	PSI1	2.765E -02	2.747E -02	2.714E -02	2.704E -02	2.697E -02
	STU	3.176E -02	3.166E -02	3.141E -02	3.130E -02	3.119E -02
	Average	2.779E -02	2.767E -02	2.738E -02	2.725E -02	2.714E -02
Np-237	BEN	-	6.582E -06	2.484E -05	3.221E -05	3.865E -05
	BNFL	-	3.577E -06	2.220E -05	2.864E -05	3.600E -05
	CEA	-	5.784E -06	2.109E -05	2.731E -05	3.275E -05
	ECN	-	5.877E -06	2.179E -05	2.818E -05	3.373E -05
	EDF	-	5.677E -06	2.067E -05	2.673E -05	3.204E -05
	HIT	-	6.503E -06	2.341E -05	2.997E -05	3.561E -05
	IKE1	-	5.067E -06	1.992E -05	2.619E -05	3.176E -05
	IKE2	-	-	-	-	-
	JAE	-	5.095E -06	1.956E -05	2.559E -05	3.090E -05
	PSI1	-	6.799E -06	2.483E -05	3.198E -05	3.819E -05
	STU	-	5.801E -06	2.192E -05	2.866E -05	3.468E -05
	Average	-	5.676E -06	2.202E -05	2.855E -05	3.443E -05
Pu-238	BEN	2.394E -03	2.249E -03	1.960E -03	1.855E -03	1.765E -03
	BNFL	2.395E -03	2.267E -03	1.979E -03	1.881E -03	1.785E -03
	CEA	2.449E -03	2.285E -03	2.048E -03	2.003E -03	1.982E -03
	ECN	2.386E -03	2.240E -03	2.041E -03	2.009E -03	1.994E -03
	EDF	2.395E -03	2.241E -03	2.030E -03	1.993E -03	1.976E -03
	HIT	2.408E -03	2.252E -03	2.026E -03	1.978E -03	1.953E -03
	IKE1	2.382E -03	2.232E -03	2.024E -03	1.987E -03	1.969E -03
	IKE2	2.403E -03	-	-	-	-
	JAE	2.398E -03	2.249E -03	2.057E -03	2.027E -03	2.018E -03
	PSI1	2.387E -03	2.238E -03	2.031E -03	1.993E -03	1.975E -03
	STU	2.553E -03	2.394E -03	2.177E -03	2.140E -03	2.124E -03
	Average	2.414E -03	2.265E -03	2.037E -03	1.987E -03	1.954E -03

Tab. 13: Fractional Fission Rate of Actinides, Benchmark A, (cont.)

Nuclide	Contributor	Burnup, MWd/kg 0.0	10.0	33.0	42.0	50.0
Pu-239	BEN	2.256E -01	2.056E -01	1.803E -01	1.725E -01	1.665E -01
	BNFL	2.256E -01	2.080E -01	1.814E -01	1.739E -01	1.673E -01
	CEA	2.330E -01	2.101E -01	1.821E -01	1.739E -01	1.676E -01
	ECN	2.331E -01	2.129E -01	1.865E -01	1.784E -01	1.722E -01
	EDF	2.360E -01	2.134E -01	1.831E -01	1.740E -01	1.669E -01
	HIT	2.345E -01	2.129E -01	1.852E -01	1.764E -01	1.695E -01
	IKE1	2.346E -01	2.131E -01	1.844E -01	1.755E -01	1.684E -01
	IKE2	2.336E -01	-	-	-	-
	JAE	2.333E -01	2.118E -01	1.839E -01	1.755E -01	1.689E -01
	PSI1	2.335E -01	2.127E -01	1.864E -01	1.787E -01	1.728E -01
	STU	2.354E -01	2.137E -01	1.844E -01	1.754E -01	1.684E -01
	Average	2.326E -01	2.114E -01	1.838E -01	1.754E -01	1.689E -01
Pu-240	BEN	5.368E -03	5.238E -03	4.877E -03	4.710E -03	4.551E -03
	BNFL	5.371E -03	5.211E -03	4.884E -03	4.732E -03	4.573E -03
	CEA	6.169E -03	5.919E -03	5.308E -03	5.058E -03	4.833E -03
	ECN	6.043E -03	5.805E -03	5.221E -03	4.984E -03	4.764E -03
	EDF	5.245E -03	5.036E -03	4.508E -03	4.289E -03	4.091E -03
	HIT	5.922E -03	5.700E -03	5.152E -03	4.912E -03	4.692E -03
	IKE1	5.965E -03	5.744E -03	5.181E -03	4.945E -03	4.728E -03
	IKE2	6.061E -03	-	-	-	-
	JAE	5.955E -03	5.687E -03	5.055E -03	4.796E -03	4.563E -03
	PSI1	5.992E -03	5.761E -03	5.207E -03	4.983E -03	4.782E -03
	STU	6.893E -03	6.606E -03	5.887E -03	5.590E -03	5.320E -03
	Average	5.908E -03	5.671E -03	5.128E -03	4.900E -03	4.690E -03
Pu-241	BEN	1.010E -01	1.004E -01	1.026E -01	1.031E -01	1.034E -01
	BNFL	1.010E -01	9.833E -02	1.015E -01	1.019E -01	1.026E -01
	CEA	1.018E -01	1.050E -01	1.103E -01	1.104E -01	1.100E -01
	ECN	1.018E -01	1.044E -01	1.105E -01	1.117E -01	1.120E -01
	EDF	9.910E -02	1.017E -01	1.076E -01	1.084E -01	1.084E -01
	HIT	1.027E -01	1.050E -01	1.112E -01	1.121E -01	1.123E -01
	IKE1	1.013E -01	1.031E -01	1.079E -01	1.086E -01	1.087E -01
	IKE2	1.015E -01	-	-	-	-
	JAE	1.015E -01	1.038E -01	1.095E -01	1.103E -01	1.103E -01
	PSI1	1.016E -01	1.031E -01	1.081E -01	1.090E -01	1.093E -01
	STU	1.025E -01	1.054E -01	1.121E -01	1.132E -01	1.135E -01
	Average	1.014E -01	1.030E -01	1.081E -01	1.089E -01	1.091E -01

Tab. 13: Fractional Fission Rate of Actinides, Benchmark A, (cont.)

Nuclide	Contributor	Burnup, MWd/kg 0.0	10.0	33.0	42.0	50.0
Pu-242	BEN	3.264E -03	3.111E -03	2.774E -03	2.648E -03	2.538E -03
	BNFL	3.266E -03	3.123E -03	2.792E -03	2.674E -03	2.559E -03
	CEA	3.278E -03	3.235E -03	3.165E -03	3.144E -03	3.126E -03
	ECN	3.209E -03	3.169E -03	3.101E -03	3.082E -03	3.065E -03
	EDF	3.108E -03	3.059E -03	2.976E -03	2.951E -03	2.932E -03
	HIT	3.449E -03	3.380E -03	3.269E -03	3.225E -03	3.186E -03
	IKE1	3.159E -03	3.126E -03	3.071E -03	3.056E -03	3.042E -03
	IKE2	3.223E -03	-	-	-	-
	JAE	3.280E -03	3.233E -03	3.195E -03	3.195E -03	3.198E -03
	PSI1	3.177E -03	3.139E -03	3.084E -03	3.072E -03	3.064E -03
	STU	3.720E -03	3.691E -03	3.653E -03	3.646E -03	3.640E -03
	Average	3.285E -03	3.227E -03	3.108E -03	3.069E -03	3.035E -03
Am-241	BEN	-	8.992E -05	2.260E -04	2.570E -04	2.756E -04
	BNFL	-	5.561E -05	2.114E -04	2.421E -04	2.677E -04
	CEA	-	8.058E -05	2.235E -04	2.633E -04	2.914E -04
	ECN	-	9.370E -05	2.400E -04	2.747E -04	2.965E -04
	EDF	-	8.882E -05	2.298E -04	2.637E -04	2.852E -04
	HIT	-	9.905E -05	2.591E -04	2.981E -04	3.228E -04
	IKE1	-	9.211E -05	2.353E -04	2.689E -04	2.898E -04
	IKE2	-	-	-	-	-
	JAE	-	9.310E -05	2.416E -04	2.775E -04	3.001E -04
	PSI1	-	8.771E -05	2.268E -04	2.608E -04	2.828E -04
	STU	-	1.054E -04	2.718E -04	3.112E -04	3.352E -04
	Average	-	8.860E -05	2.365E -04	2.717E -04	2.947E -04
Am-242m	BEN	-	1.324E -04	7.486E -04	9.575E -04	1.107E -03
	BNFL	-	5.314E -05	6.649E -04	8.604E -04	1.048E -03
	CEA	-	1.216E -04	7.138E -04	9.324E -04	1.101E -03
	ECN	-	1.245E -04	6.903E -04	8.804E -04	1.015E -03
	EDF	-	1.385E -04	7.950E -04	1.021E -03	1.185E -03
	HIT	-	-	-	-	-
	IKE1	-	1.222E -04	6.828E -04	8.707E -04	1.007E -03
	IKE2	-	-	-	-	-
	JAE	-	1.783E -04	1.041E -03	1.345E -03	1.569E -03
	PSI1	-	-	-	-	-
	STU	-	1.362E -04	7.684E -04	9.821E -04	1.137E -03
	Average	-	1.258E -04	7.631E -04	9.812E -04	1.146E -03

Tab. 13: Fractional Fission Rate of Actinides, Benchmark A, (cont.)

Nuclide	Contributor	Burnup, MWd/kg				
		0.0	10.0	33.0	42.0	50.0
Am-243	BEN	-	2.262E -04	5.969E -04	6.999E -04	7.758E -04
	BNFL	-	1.398E -04	5.532E -04	6.499E -04	7.436E -04
	CEA	-	1.309E -04	3.490E -04	4.126E -04	4.615E -04
	ECN	-	1.251E -04	3.361E -04	3.980E -04	4.454E -04
	EDF	-	1.296E -04	3.470E -04	4.106E -04	4.593E -04
	HIT	-	1.645E -04	4.409E -04	5.204E -04	5.812E -04
	IKE1	-	1.206E -04	3.264E -04	3.872E -04	4.343E -04
	IKE2	-	-	-	-	-
	JAE	-	1.289E -04	3.492E -04	4.153E -04	4.668E -04
	PSI1	-	1.224E -04	3.354E -04	4.010E -04	4.525E -04
	STU	-	1.345E -04	3.587E -04	4.233E -04	4.723E -04
	Average	-	1.422E -04	3.993E -04	4.718E -04	5.293E -04
Cm-242	BEN	-	-	-	-	-
	BNFL	-	-	-	-	-
	CEA	-	6.618E -06	3.638E -05	4.630E -05	5.354E -05
	ECN	-	7.245E -06	3.801E -05	4.860E -05	5.664E -05
	EDF	-	1.031E -05	5.303E -05	6.733E -05	7.807E -05
	HIT	-	6.623E -06	3.555E -05	4.578E -05	5.367E -05
	IKE1	-	7.103E -06	3.706E -05	4.738E -05	5.529E -05
	IKE2	-	-	-	-	-
	JAE	-	1.037E -05	5.520E -05	7.092E -05	8.295E -05
	PSI1	-	6.508E -06	3.419E -05	4.396E -05	5.115E -05
	STU	-	8.200E -06	4.359E -05	5.594E -05	6.544E -05
	Average	-	7.871E -06	4.163E -05	5.328E -05	6.209E -05
Cm-243	BEN	-	-	-	-	-
	BNFL	-	-	-	-	-
	CEA	-	3.296E -06	5.582E -05	8.726E -05	1.159E -04
	ECN	-	2.617E -06	4.466E -05	7.099E -05	9.608E -05
	EDF	-	3.381E -06	5.658E -05	8.972E -05	1.214E -04
	HIT	-	-	-	-	-
	IKE1	-	2.749E -06	4.553E -05	7.197E -05	9.712E -05
	IKE2	-	-	-	-	-
	JAE	-	2.260E -06	4.055E -05	6.542E -05	8.968E -05
	PSI1	-	2.551E -06	4.207E -05	6.680E -05	9.023E -05
	STU	-	2.665E -06	4.561E -05	7.264E -05	9.863E -05
	Average	-	2.788E -06	4.726E -05	7.497E -05	1.013E -04

Tab. 13: Fractional Fission Rate of Actinides, Benchmark A, (cont.)

Nuclide	Contributor	Burnup, MWd/kg				
		0.0	10.0	33.0	42.0	50.0
Cm-244	BEN	-	-	-	-	-
	BNFL	-	-	-	-	-
	CEA	-	3.210E -05	2.594E -04	3.797E -04	4.942E -04
	ECN	-	-	-	-	-
	EDF	-	-	-	-	-
	HIT	-	2.790E -05	2.320E -04	3.426E -04	4.485E -04
	IKE1	-	2.828E -05	2.346E -04	3.470E -04	4.564E -04
	IKE2	-	-	-	-	-
	JAE	-	2.525E -05	2.072E -04	3.055E -04	4.004E -04
	PSI1	-	-	-	-	-
	STU	-	3.089E -05	2.578E -04	3.799E -04	4.979E -04
	Average	-	2.888E -05	2.382E -04	3.509E -04	4.595E -04
Cm-245	BEN	-	-	-	-	-
	BNFL	-	-	-	-	-
	CEA	-	-	-	-	-
	ECN	-	-	-	-	-
	EDF	-	-	-	-	-
	HIT	-	2.195E -05	5.776E -04	1.043E -03	1.573E -03
	IKE1	-	2.363E -05	6.198E -04	1.145E -03	1.762E -03
	IKE2	-	-	-	-	-
	JAE	-	2.556E -05	6.642E -04	1.206E -03	1.819E -03
	PSI1	-	-	-	-	-
	STU	-	2.639E -05	7.040E -04	1.302E -03	2.005E -03
	Average	-	2.438E -05	6.414E -04	1.174E -03	1.789E -03

Tab. 14: Neutrons per Fission, Benchmark A

Nuclide	Contributor	Burnup, MWd/kg				
		0.0	10.0	33.0	42.0	50.0
U-234	BEN	-	-	-	-	-
	BNFL	-	-	-	-	-
	CEA	-	2.636E+00	2.636E+00	2.636E+00	2.636E+00
	ECN	2.613E+00	2.614E+00	2.612E+00	2.612E+00	2.613E+00
	EDF	-	2.632E+00	2.632E+00	2.632E+00	2.632E+00
	HIT	-	2.642E+00	2.643E+00	2.643E+00	2.643E+00
	IKE1	2.632E+00	2.633E+00	2.633E+00	2.633E+00	2.633E+00
	IKE2	-	-	-	-	-
	JAE	2.633E+00	2.632E+00	2.632E+00	2.632E+00	2.632E+00
	PSI1	2.637E+00	2.637E+00	2.637E+00	2.636E+00	2.636E+00
	STU	2.352E+00	2.352E+00	2.352E+00	2.352E+00	2.352E+00
	Average	2.573E+00	2.597E+00	2.597E+00	2.597E+00	2.597E+00
U-235	BEN	2.445E+00	2.444E+00	2.442E+00	2.444E+00	2.444E+00
	BNFL	2.444E+00	2.444E+00	2.444E+00	2.443E+00	2.443E+00
	CEA	2.446E+00	2.446E+00	2.446E+00	2.446E+00	2.446E+00
	ECN	2.456E+00	2.456E+00	2.455E+00	2.456E+00	2.454E+00
	EDF	2.431E+00	2.431E+00	2.430E+00	2.430E+00	2.430E+00
	HIT	2.434E+00	2.434E+00	2.434E+00	2.434E+00	2.434E+00
	IKE1	2.446E+00	2.447E+00	2.446E+00	2.446E+00	2.446E+00
	IKE2	2.446E+00	-	-	-	-
	JAE	2.438E+00	2.438E+00	2.438E+00	2.438E+00	2.438E+00
	PSI1	2.447E+00	2.447E+00	2.446E+00	2.446E+00	2.446E+00
	STU	2.422E+00	2.422E+00	2.422E+00	2.422E+00	2.422E+00
	Average	2.441E+00	2.441E+00	2.440E+00	2.441E+00	2.440E+00
U-236	BEN	2.750E+00	2.750E+00	2.750E+00	2.750E+00	2.750E+00
	BNFL	-	2.779E+00	2.779E+00	2.779E+00	2.779E+00
	CEA	-	2.580E+00	2.585E+00	2.587E+00	2.588E+00
	ECN	2.557E+00	2.561E+00	2.565E+00	2.566E+00	2.566E+00
	EDF	-	2.575E+00	2.581E+00	2.583E+00	2.584E+00
	HIT	-	2.647E+00	2.655E+00	2.656E+00	2.658E+00
	IKE1	2.566E+00	2.569E+00	2.573E+00	2.574E+00	2.574E+00
	IKE2	-	-	-	-	-
	JAE	2.640E+00	2.636E+00	2.641E+00	2.642E+00	2.643E+00
	PSI1	2.579E+00	2.579E+00	2.585E+00	2.587E+00	2.588E+00
	STU	2.317E+00	2.317E+00	2.317E+00	2.317E+00	2.317E+00
	Average	2.568E+00	2.599E+00	2.603E+00	2.604E+00	2.605E+00

Tab. 14: Neutrons per Fission, Benchmark A, (cont.)

Nuclide	Contributor	Burnup, MWd/kg				
		0.0	10.0	33.0	42.0	50.0
U-238	BEN	2.917E+00	2.917E+00	2.917E+00	2.917E+00	2.917E+00
	BNFL	2.802E+00	2.802E+00	2.802E+00	2.802E+00	2.802E+00
	CEA	2.801E+00	2.805E+00	2.807E+00	2.806E+00	2.806E+00
	ECN	2.734E+00	2.736E+00	2.735E+00	2.734E+00	2.735E+00
	EDF	2.814E+00	2.814E+00	2.814E+00	2.814E+00	2.814E+00
	HIT	2.818E+00	2.815E+00	2.816E+00	2.816E+00	2.816E+00
	IKE1	2.810E+00	2.810E+00	2.810E+00	2.810E+00	2.810E+00
	IKE2	2.808E+00	-	-	-	-
	JAE	2.792E+00	2.791E+00	2.791E+00	2.791E+00	2.791E+00
	PSI1	2.821E+00	2.821E+00	2.820E+00	2.820E+00	2.820E+00
	STU	2.441E+00	2.441E+00	2.441E+00	2.441E+00	2.441E+00
	Average	2.778E+00	2.775E+00	2.775E+00	2.775E+00	2.775E+00
Np-237	BEN	2.881E+00	2.881E+00	2.881E+00	2.881E+00	2.881E+00
	BNFL	-	2.868E+00	2.869E+00	2.869E+00	2.869E+00
	CEA	-	2.879E+00	2.879E+00	2.879E+00	2.879E+00
	ECN	2.856E+00	2.857E+00	2.857E+00	2.856E+00	2.856E+00
	EDF	2.874E+00	2.874E+00	2.874E+00	2.874E+00	2.875E+00
	HIT	-	2.839E+00	2.840E+00	2.839E+00	2.839E+00
	IKE1	2.875E+00	2.875E+00	2.875E+00	2.875E+00	2.875E+00
	IKE2	-	-	-	-	-
	JAE	2.857E+00	2.856E+00	2.856E+00	2.856E+00	2.855E+00
	PSI1	2.881E+00	2.881E+00	2.880E+00	2.880E+00	2.879E+00
	STU	2.534E+00	2.534E+00	2.534E+00	2.534E+00	2.534E+00
	Average	2.823E+00	2.834E+00	2.834E+00	2.834E+00	2.834E+00
Pu-238	BEN	3.038E+00	3.038E+00	3.043E+00	3.037E+00	3.043E+00
	BNFL	3.041E+00	3.041E+00	3.040E+00	3.040E+00	3.040E+00
	CEA	3.043E+00	3.045E+00	3.043E+00	3.042E+00	3.041E+00
	ECN	3.044E+00	3.043E+00	3.043E+00	3.042E+00	3.041E+00
	EDF	3.040E+00	3.040E+00	3.039E+00	3.039E+00	3.039E+00
	HIT	3.045E+00	3.044E+00	3.044E+00	3.043E+00	3.042E+00
	IKE1	3.041E+00	3.041E+00	3.040E+00	3.040E+00	3.040E+00
	IKE2	3.044E+00	-	-	-	-
	JAE	3.047E+00	3.047E+00	3.045E+00	3.044E+00	3.044E+00
	PSI1	3.043E+00	3.043E+00	3.042E+00	3.041E+00	3.041E+00
	STU	2.895E+00	2.895E+00	2.895E+00	2.895E+00	2.895E+00
	Average	3.029E+00	3.028E+00	3.027E+00	3.026E+00	3.027E+00

Tab. 14: Neutrons per Fission, Benchmark A, (cont.)

Nuclide	Contributor	Burnup, MWd/kg				
		0.0	10.0	33.0	42.0	50.0
Pu-239	BEN	2.881E+00	2.881E+00	2.881E+00	2.881E+00	2.881E+00
	BNFL	2.882E+00	2.882E+00	2.881E+00	2.880E+00	2.880E+00
	CEA	2.875E+00	2.879E+00	2.878E+00	2.878E+00	2.878E+00
	ECN	2.890E+00	2.889E+00	2.887E+00	2.888E+00	2.887E+00
	EDF	2.859E+00	2.859E+00	2.858E+00	2.858E+00	2.858E+00
	HIT	2.885E+00	2.885E+00	2.884E+00	2.884E+00	2.883E+00
	IKE1	2.872E+00	2.872E+00	2.871E+00	2.871E+00	2.871E+00
	IKE2	2.880E+00	-	-	-	-
	JAE	2.885E+00	2.885E+00	2.884E+00	2.883E+00	2.883E+00
	PSI1	2.872E+00	2.872E+00	2.872E+00	2.872E+00	2.872E+00
	STU	2.860E+00	2.860E+00	2.860E+00	2.860E+00	2.860E+00
	Average	2.876E+00	2.876E+00	2.876E+00	2.875E+00	2.875E+00
Pu-240	BEN	3.150E+00	3.167E+00	3.167E+00	3.167E+00	3.150E+00
	BNFL	3.152E+00	3.152E+00	3.152E+00	3.152E+00	3.152E+00
	CEA	3.086E+00	3.087E+00	3.082E+00	3.083E+00	3.083E+00
	ECN	3.069E+00	3.068E+00	3.067E+00	3.068E+00	3.068E+00
	EDF	3.126E+00	3.126E+00	3.126E+00	3.126E+00	3.126E+00
	HIT	3.159E+00	3.158E+00	3.158E+00	3.158E+00	3.157E+00
	IKE1	3.080E+00	3.080E+00	3.079E+00	3.079E+00	3.079E+00
	IKE2	3.085E+00	-	-	-	-
	JAE	3.094E+00	3.093E+00	3.092E+00	3.092E+00	3.092E+00
	PSI1	3.085E+00	3.085E+00	3.084E+00	3.083E+00	3.083E+00
	STU	2.775E+00	2.775E+00	2.775E+00	2.775E+00	2.775E+00
	Average	3.078E+00	3.079E+00	3.078E+00	3.078E+00	3.077E+00
Pu-241	BEN	2.940E+00	2.939E+00	2.939E+00	2.940E+00	2.939E+00
	BNFL	2.940E+00	2.940E+00	2.940E+00	2.940E+00	2.939E+00
	CEA	2.940E+00	2.940E+00	2.940E+00	2.940E+00	2.939E+00
	ECN	2.952E+00	2.953E+00	2.951E+00	2.951E+00	2.951E+00
	EDF	2.968E+00	2.968E+00	2.968E+00	2.967E+00	2.967E+00
	HIT	2.940E+00	2.940E+00	2.940E+00	2.940E+00	2.940E+00
	IKE1	2.940E+00	2.940E+00	2.940E+00	2.939E+00	2.939E+00
	IKE2	2.940E+00	-	-	-	-
	JAE	2.940E+00	2.940E+00	2.939E+00	2.939E+00	2.939E+00
	PSI1	2.940E+00	2.940E+00	2.940E+00	2.940E+00	2.939E+00
	STU	2.917E+00	2.917E+00	2.917E+00	2.917E+00	2.917E+00
	Average	2.941E+00	2.942E+00	2.941E+00	2.941E+00	2.941E+00

Tab. 14: Neutrons per Fission, Benchmark A, (cont.)

Nuclide	Contributor	Burnup, MWd/kg				
		0.0	10.0	33.0	42.0	50.0
Pu-242	BEN	3.077E+00	3.096E+00	3.096E+00	3.096E+00	3.096E+00
	BNFL	3.112E+00	3.112E+00	3.112E+00	3.112E+00	3.112E+00
	CEA	3.128E+00	3.126E+00	3.128E+00	3.127E+00	3.126E+00
	ECN	3.111E+00	3.112E+00	3.111E+00	3.110E+00	3.111E+00
	EDF	3.159E+00	3.159E+00	3.159E+00	3.159E+00	3.159E+00
	HIT	3.144E+00	3.142E+00	3.142E+00	3.142E+00	3.142E+00
	IKE1	3.120E+00	3.121E+00	3.121E+00	3.121E+00	3.121E+00
	IKE2	3.125E+00	-	-	-	-
	JAE	3.124E+00	3.123E+00	3.122E+00	3.122E+00	3.122E+00
	PSI1	3.126E+00	3.126E+00	3.125E+00	3.125E+00	3.125E+00
	STU	2.808E+00	2.808E+00	2.808E+00	2.808E+00	2.808E+00
	Average	3.094E+00	3.093E+00	3.092E+00	3.092E+00	3.092E+00
Am-241	BEN	3.603E+00	3.603E+00	3.615E+00	3.582E+00	3.608E+00
	BNFL	-	3.611E+00	3.610E+00	3.609E+00	3.607E+00
	CEA	-	3.614E+00	3.614E+00	3.613E+00	3.612E+00
	ECN	3.607E+00	3.608E+00	3.608E+00	3.606E+00	3.604E+00
	EDF	3.619E+00	3.622E+00	3.622E+00	3.622E+00	3.621E+00
	HIT	-	3.510E+00	3.510E+00	3.509E+00	3.508E+00
	IKE1	3.607E+00	3.609E+00	3.609E+00	3.608E+00	3.607E+00
	IKE2	-	-	-	-	-
	JAE	3.506E+00	3.508E+00	3.507E+00	3.506E+00	3.505E+00
	PSI1	3.619E+00	3.619E+00	3.618E+00	3.616E+00	3.615E+00
	STU	3.330E+00	3.330E+00	3.330E+00	3.330E+00	3.330E+00
	Average	3.556E+00	3.563E+00	3.564E+00	3.560E+00	3.562E+00
Am-242m	BEN	2.702E+00	2.702E+00	2.702E+00	2.702E+00	2.702E+00
	BNFL	-	2.702E+00	2.702E+00	2.702E+00	2.702E+00
	CEA	-	3.212E+00	3.212E+00	3.212E+00	3.212E+00
	ECN	-	-	-	-	-
	EDF	3.253E+00	3.253E+00	3.253E+00	3.253E+00	3.253E+00
	HIT	-	-	-	-	-
	IKE1	3.212E+00	3.212E+00	3.212E+00	3.212E+00	3.212E+00
	IKE2	-	-	-	-	-
	JAE	3.277E+00	3.277E+00	3.277E+00	3.277E+00	3.277E+00
	PSI1	-	-	-	-	-
	STU	3.210E+00	3.210E+00	3.210E+00	3.210E+00	3.210E+00
	Average	3.131E+00	3.081E+00	3.081E+00	3.081E+00	3.081E+00

Tab. 14: Neutrons per Fission, Benchmark A, (cont.)

Nuclide	Contributor	Burnup, MWd/kg 0.0	10.0	33.0	42.0	50.0
Am-243	BEN	3.469E+00	3.449E+00	3.521E+00	3.521E+00	3.521E+00
	BNFL	-	3.485E+00	3.487E+00	3.488E+00	3.488E+00
	CEA	-	3.497E+00	3.499E+00	3.500E+00	3.500E+00
	ECN	3.479E+00	3.481E+00	3.481E+00	3.482E+00	3.481E+00
	EDF	3.490E+00	3.492E+00	3.494E+00	3.494E+00	3.495E+00
	HIT	-	3.553E+00	3.559E+00	3.559E+00	3.560E+00
	IKE1	3.492E+00	3.493E+00	3.495E+00	3.496E+00	3.496E+00
	IKE2	-	-	-	-	-
	JAE	3.580E+00	3.582E+00	3.583E+00	3.583E+00	3.583E+00
	PSI1	3.501E+00	3.501E+00	3.503E+00	3.503E+00	3.503E+00
	STU	3.064E+00	3.064E+00	3.064E+00	3.064E+00	3.064E+00
	Average	3.439E+00	3.460E+00	3.469E+00	3.469E+00	3.469E+00
Cm-242	BEN	-	-	-	-	-
	BNFL	-	-	-	-	-
	CEA	-	3.443E+00	3.442E+00	3.441E+00	3.441E+00
	ECN	3.429E+00	3.430E+00	3.427E+00	3.427E+00	3.426E+00
	EDF	3.435E+00	3.437E+00	3.441E+00	3.442E+00	3.443E+00
	HIT	-	3.811E+00	3.811E+00	3.809E+00	3.809E+00
	IKE1	3.439E+00	3.440E+00	3.439E+00	3.438E+00	3.437E+00
	IKE2	-	-	-	-	-
	JAE	3.491E+00	3.490E+00	3.489E+00	3.488E+00	3.488E+00
	PSI1	3.444E+00	3.444E+00	3.442E+00	3.442E+00	3.441E+00
	STU	3.150E+00	3.150E+00	3.150E+00	3.150E+00	3.150E+00
	Average	3.398E+00	3.456E+00	3.455E+00	3.455E+00	3.454E+00
Cm-243	BEN	-	-	-	-	-
	BNFL	-	-	-	-	-
	CEA	-	3.397E+00	3.397E+00	3.397E+00	3.397E+00
	ECN	3.427E+00	3.428E+00	3.429E+00	3.427E+00	3.427E+00
	EDF	3.437E+00	3.437E+00	3.437E+00	3.437E+00	3.437E+00
	HIT	-	-	-	-	-
	IKE1	3.397E+00	3.397E+00	3.398E+00	3.398E+00	3.398E+00
	IKE2	-	-	-	-	-
	JAE	3.441E+00	3.441E+00	3.441E+00	3.441E+00	3.441E+00
	PSI1	3.398E+00	3.398E+00	3.398E+00	3.398E+00	3.398E+00
	STU	3.390E+00	3.390E+00	3.390E+00	3.390E+00	3.390E+00
	Average	3.415E+00	3.413E+00	3.413E+00	3.412E+00	3.412E+00

Tab. 14: Neutrons per Fission, Benchmark A, (cont.)

Nuclide	Contributor	Burnup, MWd/kg 0.0	10.0	33.0	42.0	50.0
Cm-244	BEN	-	-	-	-	-
	BNFL	-	-	-	-	-
	CEA	-	3.539E+00	3.543E+00	3.544E+00	3.546E+00
	ECN	-	-	-	-	-
	EDF	-	-	-	-	-
	HIT	-	3.523E+00	3.531E+00	3.532E+00	3.533E+00
	IKE1	3.538E+00	3.539E+00	3.541E+00	3.542E+00	3.542E+00
	IKE2	-	-	-	-	-
	JAE	3.575E+00	3.571E+00	3.572E+00	3.572E+00	3.572E+00
	PSI1	-	-	-	-	-
	STU	3.240E+00	3.240E+00	3.240E+00	3.240E+00	3.240E+00
	Average	3.451E+00	3.483E+00	3.485E+00	3.486E+00	3.487E+00
Cm-245	BEN	-	-	-	-	-
	BNFL	-	-	-	-	-
	CEA	-	-	-	-	-
	ECN	-	-	-	-	-
	EDF	-	-	-	-	-
	HIT	-	3.840E+00	3.839E+00	3.839E+00	3.839E+00
	IKE1	3.830E+00	3.830E+00	3.830E+00	3.830E+00	3.830E+00
	IKE2	-	-	-	-	-
	JAE	3.610E+00	3.611E+00	3.610E+00	3.610E+00	3.610E+00
	PSI1	-	-	-	-	-
	STU	3.820E+00	3.820E+00	3.820E+00	3.820E+00	3.820E+00
	Average	3.754E+00	3.775E+00	3.775E+00	3.775E+00	3.775E+00

Tab. 15: Nuclide Densities of Actinides, Benchmark A

Nuclide	Contributor	Burnup, MWd/kg 0.0	10.0	33.0	42.0	50.0
U-234	BEN	-	-	-	-	-
	BNFL	-	-	-	-	-
	CEA	-	5.041E -07	1.474E -06	1.799E -06	2.070E -06
	ECN	-	5.966E -07	1.618E -06	1.919E -06	2.151E -06
	EDF	-	5.931E -07	1.595E -06	1.872E -06	2.071E -06
	HIT	-	5.977E -07	1.650E -06	1.973E -06	2.229E -06
	IKE1	-	5.940E -07	1.623E -06	1.928E -06	2.162E -06
	JAE	-	5.990E -07	1.649E -06	1.969E -06	2.223E -06
	PSI1	-	5.710E -07	1.546E -06	1.834E -06	2.056E -06
	STU	-	5.966E -07	1.641E -06	1.948E -06	2.186E -06
	Average	-	5.815E -07	1.600E -06	1.905E -06	2.144E -06
U-235	BEN	1.446E -04	1.293E -04	9.792E -05	8.701E -05	7.794E -05
	BNFL	1.446E -04	1.293E -04	9.793E -05	8.702E -05	7.796E -05
	CEA	1.446E -04	1.293E -04	9.812E -05	8.739E -05	7.851E -05
	ECN	1.445E -04	1.297E -04	9.919E -05	8.865E -05	7.997E -05
	EDF	1.446E -04	1.295E -04	9.844E -05	8.773E -05	7.887E -05
	HIT	1.446E -04	1.292E -04	9.789E -05	8.715E -05	7.829E -05
	IKE1	1.446E -04	1.295E -04	9.888E -05	8.831E -05	7.958E -05
	JAE	1.446E -04	1.292E -04	9.775E -05	8.695E -05	7.804E -05
	PSI1	1.446E -04	1.295E -04	9.869E -05	8.810E -05	7.937E -05
	STU	1.446E -04	1.297E -04	9.930E -05	8.881E -05	8.011E -05
	Average	1.446E -04	1.294E -04	9.841E -05	8.771E -05	7.886E -05
U-236	BEN	-	4.041E -06	1.152E -05	1.379E -05	1.552E -05
	BNFL	-	4.040E -06	1.152E -05	1.379E -05	1.552E -05
	CEA	-	3.831E -06	1.084E -05	1.296E -05	1.458E -05
	ECN	-	3.807E -06	1.083E -05	1.298E -05	1.462E -05
	EDF	-	3.907E -06	1.108E -05	1.325E -05	1.491E -05
	HIT	-	3.987E -06	1.136E -05	1.361E -05	1.534E -05
	IKE1	-	3.947E -06	1.116E -05	1.332E -05	1.496E -05
	JAE	-	4.240E -06	1.198E -05	1.430E -05	1.605E -05
	PSI1	-	3.900E -06	1.108E -05	1.328E -05	1.495E -05
	STU	-	3.780E -06	1.070E -05	1.280E -05	1.439E -05
	Average	-	3.948E -06	1.121E -05	1.341E -05	1.508E -05

Tab. 15: Nuclide Densities of Actinides, Benchmark A, (cont.)

Nuclide	Contributor	Burnup, MWd/kg 0.0	10.0	33.0	42.0	50.0
U-238	BEN	1.994E -02	1.981E -02	1.949E -02	1.936E -02	1.924E -02
	BNFL	1.994E -02	1.981E -02	1.949E -02	1.936E -02	1.924E -02
	CEA	1.994E -02	1.981E -02	1.949E -02	1.936E -02	1.924E -02
	ECN	1.993E -02	1.981E -02	1.949E -02	1.937E -02	1.925E -02
	EDF	1.994E -02	1.981E -02	1.950E -02	1.938E -02	1.926E -02
	HIT	1.994E -02	1.981E -02	1.950E -02	1.938E -02	1.926E -02
	IKE1	1.994E -02	1.981E -02	1.951E -02	1.938E -02	1.927E -02
	JAE	1.994E -02	1.981E -02	1.950E -02	1.938E -02	1.926E -02
	PSI1	1.994E -02	1.981E -02	1.949E -02	1.937E -02	1.925E -02
	STU	1.994E -02	1.981E -02	1.951E -02	1.939E -02	1.928E -02
	Average	1.994E -02	1.981E -02	1.950E -02	1.937E -02	1.926E -02
Np-237	BEN	-	9.864E -07	3.700E -06	4.785E -06	5.729E -06
	BNFL	-	9.867E -07	3.702E -06	4.787E -06	5.731E -06
	CEA	-	8.581E -07	3.117E -06	4.028E -06	4.821E -06
	ECN	-	8.901E -07	3.283E -06	4.235E -06	5.061E -06
	EDF	-	8.742E -07	3.168E -06	4.090E -06	4.892E -06
	HIT	-	9.749E -07	3.495E -06	4.476E -06	5.323E -06
	IKE1	-	7.789E -07	3.049E -06	4.001E -06	4.843E -06
	JAE	-	7.319E -07	2.812E -06	3.681E -06	4.445E -06
	PSI1	-	1.041E -06	3.778E -06	4.846E -06	5.766E -06
	STU	-	7.487E -07	2.811E -06	3.667E -06	4.427E -06
	Average	-	8.870E -07	3.291E -06	4.260E -06	5.104E -06
Pu-238	BEN	1.147E -04	1.077E -04	9.267E -05	8.716E -05	8.243E -05
	BNFL	1.147E -04	1.077E -04	9.268E -05	8.716E -05	8.244E -05
	CEA	1.147E -04	1.071E -04	9.492E -05	9.234E -05	9.091E -05
	ECN	1.147E -04	1.075E -04	9.665E -05	9.449E -05	9.328E -05
	EDF	1.147E -04	1.074E -04	9.642E -05	9.423E -05	9.303E -05
	HIT	1.147E -04	1.072E -04	9.546E -05	9.291E -05	9.139E -05
	IKE1	1.147E -04	1.075E -04	9.665E -05	9.444E -05	9.321E -05
	JAE	1.147E -04	1.077E -04	9.759E -05	9.581E -05	9.502E -05
	PSI1	1.147E -04	1.075E -04	9.625E -05	9.373E -05	9.217E -05
	STU	1.147E -04	1.075E -04	9.667E -05	9.454E -05	9.339E -05
	Average	1.147E -04	1.075E -04	9.560E -05	9.268E -05	9.073E -05

Tab. 15: Nuclide Densities of Actinides, Benchmark A, (cont.)

Nuclide	Contributor	Burnup, MWd/kg				
		0.0	10.0	33.0	42.0	50.0
	BEN	1.029E -03	9.320E -04	7.492E -04	6.891E -04	6.404E -04
	BNFL	1.029E -03	9.320E -04	7.493E -04	6.892E -04	6.405E -04
	CEA	1.029E -03	9.288E -04	7.436E -04	6.837E -04	6.355E -04
	ECN	1.029E -03	9.316E -04	7.522E -04	6.937E -04	6.474E -04
	EDF	1.028E -03	9.298E -04	7.444E -04	6.853E -04	6.381E -04
Pu-239	HIT	1.029E -03	9.302E -04	7.471E -04	6.876E -04	6.400E -04
	IKE1	1.029E -03	9.311E -04	7.468E -04	6.867E -04	6.386E -04
	JAE	1.029E -03	9.313E -04	7.504E -04	6.924E -04	6.460E -04
	PSI1	1.029E -03	9.361E -04	7.608E -04	7.046E -04	6.596E -04
	STU	1.029E -03	9.283E -04	7.392E -04	6.778E -04	6.292E -04
	Average	1.028E -03	9.311E -04	7.483E -04	6.890E -04	6.415E -04
	BEN	7.966E -04	7.754E -04	7.175E -04	6.912E -04	6.663E -04
	BNFL	7.966E -04	7.754E -04	7.172E -04	6.908E -04	6.657E -04
	CEA	7.966E -04	7.631E -04	6.807E -04	6.471E -04	6.168E -04
	ECN	7.967E -04	7.636E -04	6.823E -04	6.492E -04	6.191E -04
	EDF	7.965E -04	7.631E -04	6.797E -04	6.453E -04	6.142E -04
Pu-240	HIT	7.966E -04	7.665E -04	6.895E -04	6.572E -04	6.280E -04
	IKE1	7.966E -04	7.653E -04	6.866E -04	6.538E -04	6.240E -04
	JAE	7.966E -04	7.627E -04	6.782E -04	6.433E -04	6.119E -04
	PSI1	7.966E -04	7.650E -04	6.863E -04	6.540E -04	6.248E -04
	STU	7.966E -04	7.611E -04	6.736E -04	6.378E -04	6.054E -04
	Average	7.966E -04	7.661E -04	6.892E -04	6.570E -04	6.276E -04
	BEN	3.400E -04	3.410E -04	3.303E -04	3.226E -04	3.145E -04
	BNFL	3.400E -04	3.411E -04	3.305E -04	3.229E -04	3.149E -04
	CEA	3.400E -04	3.527E -04	3.521E -04	3.441E -04	3.344E -04
	ECN	3.399E -04	3.511E -04	3.535E -04	3.478E -04	3.404E -04
	EDF	3.399E -04	3.546E -04	3.619E -04	3.572E -04	3.504E -04
Pu-241	HIT	3.400E -04	3.512E -04	3.542E -04	3.489E -04	3.417E -04
	IKE1	3.400E -04	3.495E -04	3.499E -04	3.439E -04	3.364E -04
	JAE	3.400E -04	3.518E -04	3.548E -04	3.491E -04	3.416E -04
	PSI1	3.400E -04	3.497E -04	3.503E -04	3.445E -04	3.373E -04
	STU	3.400E -04	3.537E -04	3.590E -04	3.535E -04	3.458E -04
	Average	3.400E -04	3.496E -04	3.497E -04	3.434E -04	3.357E -04

Tab. 15: Nuclide Densities of Actinides, Benchmark A, (cont.)

Nuclide	Contributor	Burnup, MWd/kg				
		0.0	10.0	33.0	42.0	50.0
	BEN	5.639E -04	5.359E -04	4.744E -04	4.515E -04	4.317E -04
	BNFL	5.639E -04	5.359E -04	4.743E -04	4.514E -04	4.316E -04
	CEA	5.639E -04	5.557E -04	5.414E -04	5.368E -04	5.328E -04
	ECN	5.638E -04	5.557E -04	5.410E -04	5.362E -04	5.324E -04
	EDF	5.642E -04	5.543E -04	5.367E -04	5.311E -04	5.266E -04
Pu-242	HIT	5.639E -04	5.533E -04	5.329E -04	5.261E -04	5.204E -04
	IKE1	5.639E -04	5.568E -04	5.444E -04	5.406E -04	5.374E -04
	JAE	5.639E -04	5.576E -04	5.516E -04	5.516E -04	5.522E -04
	PSI1	5.639E -04	5.567E -04	5.438E -04	5.397E -04	5.362E -04
	STU	5.639E -04	5.580E -04	5.490E -04	5.467E -04	5.448E -04
	Average	5.639E -04	5.520E -04	5.290E -04	5.212E -04	5.146E -04
	BEN	-	1.030E -05	2.559E -05	2.887E -05	3.071E -05
	BNFL	-	1.030E -05	2.560E -05	2.888E -05	3.073E -05
	CEA	-	8.938E -06	2.465E -05	2.889E -05	3.179E -05
	ECN	-	1.055E -05	2.681E -05	3.050E -05	3.271E -05
	EDF	-	1.067E -05	2.752E -05	3.146E -05	3.387E -05
Am-241	HIT	-	1.071E -05	2.788E -05	3.201E -05	3.459E -05
	IKE1	-	1.053E -05	2.675E -05	3.041E -05	3.258E -05
	JAE	-	1.068E -05	2.770E -05	3.173E -05	3.422E -05
	PSI1	-	1.018E -05	2.612E -05	2.981E -05	3.207E -05
	STU	-	1.060E -05	2.705E -05	3.076E -05	3.292E -05
	Average	-	1.034E -05	2.657E -05	3.033E -05	3.262E -05
	BEN	-	1.122E -07	5.842E -07	7.161E -07	7.948E -07
	BNFL	-	1.123E -07	5.847E -07	7.169E -07	7.958E -07
	CEA	-	9.589E -08	5.255E -07	6.619E -07	7.547E -07
	ECN	-	9.804E -08	5.040E -07	6.197E -07	6.907E -07
	EDF	-	1.151E -07	6.210E -07	7.732E -07	8.710E -07
Am-242m	HIT	-	-	-	-	-
	IKE1	-	9.888E -08	5.154E -07	6.353E -07	7.102E -07
	JAE	-	1.504E -07	8.201E -07	1.026E -06	1.159E -06
	PSI1	-	-	-	-	-
	STU	-	1.083E -07	5.679E -07	6.995E -07	7.822E -07
	Average	-	1.114E -07	5.904E -07	7.310E -07	8.198E -07

Tab. 15: Nuclide Densities of Actinides, Benchmark A, (cont.)

Nuclide	Contributor	Burnup, MWd/kg				
		0.0	10.0	33.0	42.0	50.0
Am-243	BEN	-	4.141E -05	1.090E -04	1.276E -04	1.411E -04
	BNFL	-	4.143E -05	1.091E -04	1.277E -04	1.412E -04
	CEA	-	2.360E -05	6.296E -05	7.437E -05	8.309E -05
	ECN	-	2.301E -05	6.179E -05	7.305E -05	8.166E -05
	EDF	-	2.436E -05	6.524E -05	7.709E -05	8.612E -05
	HIT	-	2.351E -05	6.366E -05	7.537E -05	8.443E -05
	IKE1	-	2.260E -05	6.113E -05	7.245E -05	8.115E -05
	JAE	-	2.253E -05	6.149E -05	7.324E -05	8.240E -05
	PSI1	-	2.285E -05	6.269E -05	7.477E -05	8.412E -05
	STU	-	2.071E -05	5.504E -05	6.483E -05	7.221E -05
	Average	-	2.660E -05	7.121E -05	8.404E -05	9.375E -05
Cm-242	BEN	-	-	-	-	-
	BNFL	-	-	-	-	-
	CEA	-	6.187E -07	3.374E -06	4.277E -06	4.924E -06
	ECN	-	6.895E -07	3.582E -06	4.557E -06	5.287E -06
	EDF	-	6.892E -07	3.605E -06	4.602E -06	5.357E -06
	HIT	-	6.284E -07	3.346E -06	4.300E -06	5.032E -06
	IKE1	-	6.851E -07	3.546E -06	4.515E -06	5.248E -06
	JAE	-	7.189E -07	3.818E -06	4.897E -06	5.716E -06
	PSI1	-	6.266E -07	3.258E -06	4.163E -06	4.814E -06
	STU	-	6.964E -07	3.667E -06	4.685E -06	5.458E -06
	Average	-	6.691E -07	3.525E -06	4.499E -06	5.230E -06
Cm-243	BEN	-	-	-	-	-
	BNFL	-	-	-	-	-
	CEA	-	5.858E -09	9.990E -08	1.560E -07	2.065E -07
	ECN	-	5.617E -09	9.623E -08	1.524E -07	2.052E -07
	EDF	-	6.579E -09	1.114E -07	1.764E -07	2.381E -07
	HIT	-	-	-	-	-
	IKE1	-	5.549E -09	9.259E -08	1.462E -07	1.965E -07
	JAE	-	5.428E -09	9.727E -08	1.561E -07	2.127E -07
	PSI1	-	5.267E -09	8.716E -08	1.377E -07	1.849E -07
	STU	-	5.657E -09	9.777E -08	1.555E -07	2.104E -07
	Average	-	5.708E -09	9.747E -08	1.543E -07	2.077E -07

Tab. 15: Nuclide Densities of Actinides, Benchmark A, (cont.)

Nuclide	Contributor	Burnup, MWd/kg				
		0.0	10.0	33.0	42.0	50.0
Cm-244	BEN	-	-	-	-	-
	BNFL	-	-	-	-	-
	CEA	-	2.805E -06	2.288E -05	3.358E -05	4.381E -05
	ECN	-	-	-	-	-
	EDF	-	-	-	-	-
	HIT	-	2.569E -06	2.182E -05	3.237E -05	4.260E -05
	IKE1	-	2.585E -06	2.149E -05	3.179E -05	4.177E -05
	JAE	-	2.520E -06	2.077E -05	3.067E -05	4.026E -05
	PSI1	-	-	-	-	-
	STU	-	2.445E -06	2.042E -05	3.007E -05	3.936E -05
	Average	-	2.585E -06	2.148E -05	3.170E -05	4.156E -05
Cm-245	BEN	-	-	-	-	-
	BNFL	-	-	-	-	-
	CEA	-	-	-	-	-
	ECN	-	-	-	-	-
	EDF	-	-	-	-	-
	HIT	-	6.999E -08	1.651E -06	2.922E -06	4.328E -06
	IKE1	-	6.589E -08	1.665E -06	3.014E -06	4.538E -06
	JAE	-	7.714E -08	1.939E -06	3.456E -06	5.116E -06
	PSI1	-	-	-	-	-
	STU	-	7.529E -08	1.931E -06	3.490E -06	5.259E -06
	Average	-	7.208E -08	1.796E -06	3.221E -06	4.810E -06

Tab. 16: Nuclide Densities of Fiss. Prod., Benchmark A

Nuclide	Contributor	Burnup, MWd/kg				
		0.0	10.0	33.0	42.0	50.0
Mo-95	BEN	-	1.160E -05	3.718E -05	4.678E -05	5.513E -05
	BNFL	-	1.160E -05	3.718E -05	4.678E -05	5.513E -05
	CEA	-	4.899E -06	2.957E -05	3.902E -05	4.716E -05
	ECN	-	5.701E -06	3.070E -05	4.025E -05	4.854E -05
	EDF	-	-	-	-	-
	HIT	-	5.576E -06	2.987E -05	3.914E -05	4.717E -05
	IKE1	-	5.384E -06	2.899E -05	3.797E -05	4.571E -05
	JAE	-	1.095E -05	3.503E -05	4.406E -05	5.191E -05
	PSI1	-	5.127E -06	2.798E -05	3.660E -05	4.399E -05
	STU	-	-	-	-	-
	Average	-	7.606E -06	3.206E -05	4.133E -05	4.934E -05
Tc-99	BEN	-	1.376E -05	4.349E -05	5.441E -05	6.380E -05
	BNFL	-	1.376E -05	4.349E -05	5.441E -05	6.380E -05
	CEA	-	1.396E -05	4.424E -05	5.518E -05	6.450E -05
	ECN	-	1.173E -05	3.657E -05	4.538E -05	5.282E -05
	EDF	-	-	-	-	-
	HIT	-	1.397E -05	4.427E -05	5.528E -05	6.467E -05
	IKE1	-	1.379E -05	4.352E -05	5.420E -05	6.323E -05
	JAE	-	1.382E -05	4.383E -05	5.475E -05	6.408E -05
	PSI1	-	1.346E -05	4.238E -05	5.281E -05	6.168E -05
	STU	-	-	-	-	-
	Average	-	1.353E -05	4.272E -05	5.330E -05	6.232E -05
Ru-101	BEN	-	1.367E -05	4.433E -05	5.604E -05	6.633E -05
	BNFL	-	-	-	-	-
	CEA	-	1.439E -05	4.632E -05	5.832E -05	6.874E -05
	ECN	-	-	-	-	-
	EDF	-	-	-	-	-
	HIT	-	1.372E -05	4.415E -05	5.563E -05	6.562E -05
	IKE1	-	1.355E -05	4.369E -05	5.504E -05	6.489E -05
	JAE	-	1.362E -05	4.383E -05	5.521E -05	6.511E -05
	PSI1	-	-	-	-	-
	STU	-	-	-	-	-
	Average	-	1.379E -05	4.446E -05	5.605E -05	6.614E -05

Tab. 16: Nuclide Densities of Fiss. Prod., Benchmark A,(cont.)

Nuclide	Contributor	Burnup, MWd/kg				
		0.0	10.0	33.0	42.0	50.0
Rh-103	BEN	-	1.004E -05	3.661E -05	4.592E -05	5.367E -05
	BNFL	-	-	-	-	-
	CEA	-	1.126E -05	4.129E -05	5.127E -05	5.934E -05
	ECN	-	-	-	-	-
	EDF	-	-	-	-	-
	HIT	-	1.141E -05	4.042E -05	5.012E -05	5.801E -05
	IKE1	-	1.116E -05	3.945E -05	4.891E -05	5.657E -05
	JAE	-	1.136E -05	4.021E -05	4.987E -05	5.771E -05
	PSI1	-	-	-	-	-
	STU	-	1.499E -05	4.456E -05	5.437E -05	6.226E -05
	Average	-	1.170E -05	4.042E -05	5.008E -05	5.793E -05
Pd-105	BEN	-	1.170E -05	3.813E -05	4.823E -05	5.710E -05
	BNFL	-	-	-	-	-
	CEA	-	1.250E -05	4.079E -05	5.150E -05	6.084E -05
	ECN	-	-	-	-	-
	EDF	-	-	-	-	-
	HIT	-	1.180E -05	3.829E -05	4.830E -05	5.699E -05
	IKE1	-	1.182E -05	3.849E -05	4.859E -05	5.739E -05
	JAE	-	1.182E -05	3.851E -05	4.864E -05	5.747E -05
	PSI1	-	-	-	-	-
	STU	-	-	-	-	-
	Average	-	1.193E -05	3.884E -05	4.905E -05	5.796E -05
Pd-107	BEN	-	-	-	-	-
	BNFL	-	-	-	-	-
	CEA	-	8.188E -06	2.663E -05	3.365E -05	3.977E -05
	ECN	-	-	-	-	-
	EDF	-	-	-	-	-
	HIT	-	8.118E -06	2.639E -05	3.336E -05	3.945E -05
	IKE1	-	7.997E -06	2.596E -05	3.278E -05	3.872E -05
	JAE	-	8.226E -06	2.685E -05	3.399E -05	4.025E -05
	PSI1	-	-	-	-	-
	STU	-	-	-	-	-
	Average	-	8.132E -06	2.646E -05	3.344E -05	3.955E -05

Tab. 16: Nuclide Densities of Fiss. Prod., Benchmark A,(cont.)

Nuclide	Contributor	Burnup, MWd/kg				
		0.0	10.0	33.0	42.0	50.0
Pd-108	BEN	-	5.739E −06	1.808E −05	2.258E −05	2.643E −05
	BNFL	-	-	-	-	-
	CEA	-	5.551E −06	1.858E −05	2.374E −05	2.834E −05
	ECN	-	-	-	-	-
	EDF	-	-	-	-	-
	HIT	-	5.591E −06	1.866E −05	2.385E −05	2.848E −05
	IKE1	-	5.701E −06	1.903E −05	2.429E −05	2.896E −05
	JAE	-	5.673E −06	1.892E −05	2.419E −05	2.891E −05
	PSI1	-	-	-	-	-
	STU	-	-	-	-	-
	Average	-	5.651E −06	1.865E −05	2.373E −05	2.823E −05
Ag-109	BEN	-	3.308E −06	9.687E −06	1.179E −05	1.349E −05
	BNFL	-	3.307E −06	9.686E −06	1.179E −05	1.349E −05
	CEA	-	3.642E −06	1.026E −05	1.236E −05	1.406E −05
	ECN	-	1.831E −06	4.922E −06	5.825E −06	6.522E −06
	EDF	-	-	-	-	-
	HIT	-	3.477E −06	1.014E −05	1.238E −05	1.422E −05
	IKE1	-	3.817E −06	1.072E −05	1.287E −05	1.457E −05
	JAE	-	3.821E −06	1.083E −05	1.307E −05	1.488E −05
	PSI1	-	3.598E −06	1.028E −05	1.250E −05	1.434E −05
	STU	-	3.504E −06	9.265E −06	1.083E −05	1.195E −05
	Average	-	3.367E −06	9.532E −06	1.149E −05	1.306E −05
Xe-131	BEN	-	7.613E −06	2.145E −05	2.562E −05	2.879E −05
	BNFL	-	7.613E −06	2.145E −05	2.562E −05	2.879E −05
	CEA	-	7.569E −06	2.177E −05	2.591E −05	2.903E −05
	ECN	-	6.167E −06	1.768E −05	2.112E −05	2.375E −05
	EDF	-	-	-	-	-
	HIT	-	7.445E −06	2.211E −05	2.676E −05	3.043E −05
	IKE1	-	7.154E −06	2.027E −05	2.401E −05	2.677E −05
	JAE	-	7.242E −06	2.166E −05	2.640E −05	3.026E −05
	PSI1	-	7.737E −06	2.195E −05	2.636E −05	2.980E −05
	STU	-	7.681E −06	2.110E −05	2.493E −05	2.775E −05
	Average	-	7.358E −06	2.105E −05	2.519E −05	2.838E −05

Tab. 16: Nuclide Densities of Fiss. Prod., Benchmark A,(cont.)

Nuclide	Contributor	Burnup, MWd/kg				
		0.0	10.0	33.0	42.0	50.0
Xe-135	BEN	-	2.518E −08	2.398E −08	2.331E −08	2.284E −08
	BNFL	-	-	-	-	-
	CEA	-	2.895E −08	2.481E −08	2.336E −08	2.213E −08
	ECN	-	-	-	-	-
	EDF	-	-	-	-	-
	HIT	-	2.585E −08	2.460E −08	2.407E −08	2.358E −08
	IKE1	-	2.640E −08	2.508E −08	2.448E −08	2.391E −08
	JAE	-	2.621E −08	2.500E −08	2.448E −08	2.399E −08
	PSI1	-	-	-	-	-
	STU	-	2.549E −08	2.425E −08	2.359E −08	2.315E −08
	Average	-	2.635E −08	2.462E −08	2.388E −08	2.327E −08
Cs-133	BEN	-	1.540E −05	4.727E −05	5.845E −05	6.780E −05
	BNFL	-	1.540E −05	4.727E −05	5.845E −05	6.780E −05
	CEA	-	1.533E −05	4.798E −05	5.934E −05	6.882E −05
	ECN	-	1.292E −05	4.005E −05	4.941E −05	5.701E −05
	EDF	-	-	-	-	-
	HIT	-	1.500E −05	4.744E −05	5.894E −05	6.861E −05
	IKE1	-	1.495E −05	4.658E −05	5.754E −05	6.661E −05
	JAE	-	1.512E −05	4.783E −05	5.955E −05	6.949E −05
	PSI1	-	1.475E −05	4.622E −05	5.728E −05	6.655E −05
	STU	-	1.527E −05	4.633E −05	5.701E −05	6.581E −05
	Average	-	1.490E −05	4.633E −05	5.733E −05	6.650E −05
Cs-135	BEN	-	1.220E −05	3.865E −05	4.835E −05	5.666E −05
	BNFL	-	-	-	-	-
	CEA	-	1.205E −05	3.917E −05	4.944E −05	5.839E −05
	ECN	-	-	-	-	-
	EDF	-	-	-	-	-
	HIT	-	1.236E −05	3.945E −05	4.944E −05	5.801E −05
	IKE1	-	1.251E −05	3.984E −05	4.990E −05	5.852E −05
	JAE	-	1.256E −05	4.016E −05	5.042E −05	5.927E −05
	PSI1	-	-	-	-	-
	STU	-	1.217E −05	3.863E −05	4.829E −05	5.653E −05
	Average	-	1.231E −05	3.932E −05	4.931E −05	5.790E −05

Tab. 16: Nuclide Densities of Fiss. Prod., Benchmark A,(cont.)

Nuclide	Contributor	Burnup, MWd/kg				
		0.0	10.0	33.0	42.0	50.0
Nd-143	BEN	-	1.060E -05	3.358E -05	4.200E -05	4.920E -05
	BNFL	-	1.060E -05	3.358E -05	4.200E -05	4.921E -05
	CEA	-	9.751E -06	3.263E -05	4.081E -05	4.769E -05
	ECN	-	1.060E -05	3.556E -05	4.459E -05	5.224E -05
	EDF	-	-	-	-	-
	HIT	-	1.368E -05	4.669E -05	5.895E -05	6.946E -05
	IKE1	-	9.554E -06	3.188E -05	3.991E -05	4.668E -05
	JAE	-	9.700E -06	3.218E -05	4.029E -05	4.715E -05
	PSI1	-	9.502E -06	3.154E -05	3.947E -05	4.617E -05
	STU	-	1.036E -05	3.240E -05	4.028E -05	4.689E -05
	Average	-	1.048E -05	3.445E -05	4.314E -05	5.052E -05
Nd-145	BEN	-	-	-	-	-
	BNFL	-	7.268E -06	2.279E -05	2.842E -05	3.321E -05
	CEA	-	7.428E -06	2.332E -05	2.907E -05	3.396E -05
	ECN	-	8.455E -06	2.685E -05	3.358E -05	3.933E -05
	EDF	-	-	-	-	-
	HIT	-	7.382E -06	2.345E -05	2.938E -05	3.448E -05
	IKE1	-	7.250E -06	2.280E -05	2.844E -05	3.323E -05
	JAE	-	7.314E -06	2.315E -05	2.896E -05	3.395E -05
	PSI1	-	7.238E -06	2.282E -05	2.850E -05	3.337E -05
	STU	-	7.259E -06	2.274E -05	2.833E -05	3.305E -05
	Average	-	7.449E -06	2.349E -05	2.933E -05	3.432E -05
Pm-147	BEN	-	1.807E -06	3.458E -06	3.659E -06	3.744E -06
	BNFL	-	1.807E -06	3.458E -06	3.659E -06	3.744E -06
	CEA	-	3.799E -06	8.170E -06	8.803E -06	9.091E -06
	ECN	-	4.038E -06	8.762E -06	9.569E -06	1.003E -05
	EDF	-	-	-	-	-
	HIT	-	3.876E -06	8.563E -06	9.420E -06	9.918E -06
	IKE1	-	3.739E -06	7.927E -06	8.571E -06	8.902E -06
	JAE	-	3.798E -06	8.383E -06	9.206E -06	9.686E -06
	PSI1	-	3.736E -06	8.142E -06	8.940E -06	9.410E -06
	STU	-	3.959E -06	7.943E -06	8.531E -06	8.811E -06
	Average	-	3.396E -06	7.201E -06	7.817E -06	8.149E -06

Tab. 16: Nuclide Densities of Fiss. Prod., Benchmark A,(cont.)

Nuclide	Contributor	Burnup, MWd/kg				
		0.0	10.0	33.0	42.0	50.0
Pm-148m	BEN	-	5.818E -08	1.180E -07	1.254E -07	1.286E -07
	BNFL	-	5.820E -08	1.180E -07	1.255E -07	1.287E -07
	CEA	-	7.421E -08	1.568E -07	1.622E -07	1.615E -07
	ECN	-	6.715E -08	1.500E -07	1.625E -07	1.678E -07
	EDF	-	-	-	-	-
	HIT	-	6.480E -08	1.467E -07	1.598E -07	1.677E -07
	IKE1	-	6.890E -08	1.597E -07	1.724E -07	1.787E -07
	JAE	-	7.592E -08	1.672E -07	1.819E -07	1.886E -07
	PSI1	-	6.947E -08	1.593E -07	1.700E -07	1.750E -07
	STU	-	7.615E -08	1.670E -07	1.783E -07	1.846E -07
	Average	-	6.811E -08	1.492E -07	1.598E -07	1.646E -07
Sm-149	BEN	-	7.932E -07	8.205E -07	7.871E -07	7.476E -07
	BNFL	-	7.939E -07	8.214E -07	7.881E -07	7.486E -07
	CEA	-	5.472E -07	5.387E -07	5.160E -07	4.929E -07
	ECN	-	4.727E -07	4.719E -07	4.592E -07	4.444E -07
	EDF	-	-	-	-	-
	HIT	-	5.579E -07	5.487E -07	5.317E -07	5.141E -07
	IKE1	-	5.657E -07	5.634E -07	5.464E -07	5.286E -07
	JAE	-	5.585E -07	5.526E -07	5.389E -07	5.228E -07
	PSI1	-	5.590E -07	5.611E -07	5.470E -07	5.307E -07
	STU	-	5.564E -07	5.439E -07	5.197E -07	5.015E -07
	Average	-	6.005E -07	6.025E -07	5.816E -07	5.590E -07
Sm-150	BEN	-	2.515E -06	1.105E -05	1.445E -05	1.741E -05
	BNFL	-	2.514E -06	1.105E -05	1.445E -05	1.741E -05
	CEA	-	2.653E -06	1.047E -05	1.346E -05	1.604E -05
	ECN	-	2.218E -06	8.762E -06	1.130E -05	1.351E -05
	EDF	-	-	-	-	-
	HIT	-	2.645E -06	1.053E -05	1.365E -05	1.639E -05
	IKE1	-	2.565E -06	1.018E -05	1.314E -05	1.571E -05
	JAE	-	2.578E -06	1.012E -05	1.306E -05	1.565E -05
	PSI1	-	2.599E -06	1.064E -05	1.390E -05	1.680E -05
	STU	-	2.572E -06	9.930E -06	1.268E -05	1.501E -05
	Average	-	2.540E -06	1.030E -05	1.334E -05	1.599E -05

Tab. 16: Nuclide Densities of Fiss. Prod., Benchmark A,(cont.)

Nuclide	Contributor	Burnup, MWd/kg				
		0.0	10.0	33.0	42.0	50.0
Sm-151	BEN	-	1.560E -06	3.583E -06	4.021E -06	4.304E -06
	BNFL	-	1.560E -06	3.585E -06	4.023E -06	4.306E -06
	CEA	-	1.271E -06	2.243E -06	2.383E -06	2.466E -06
	ECN	-	9.178E -07	1.621E -06	1.729E -06	1.802E -06
	EDF	-	-	-	-	-
	HIT	-	1.279E -06	2.175E -06	2.274E -06	2.325E -06
	IKE1	-	1.277E -06	2.260E -06	2.404E -06	2.495E -06
	JAE	-	1.275E -06	2.240E -06	2.368E -06	2.440E -06
	PSI1	-	1.204E -06	1.967E -06	2.033E -06	2.063E -06
	STU	-	1.249E -06	2.247E -06	2.400E -06	2.498E -06
	Average	-	1.288E -06	2.436E -06	2.626E -06	2.744E -06
Sm-152	BEN	-	1.513E -06	4.966E -06	6.128E -06	7.054E -06
	BNFL	-	1.513E -06	4.965E -06	6.127E -06	7.052E -06
	CEA	-	1.736E -06	5.530E -06	6.646E -06	7.498E -06
	ECN	-	1.183E -06	3.723E -06	4.444E -06	4.985E -06
	EDF	-	-	-	-	-
	HIT	-	1.769E -06	6.110E -06	7.511E -06	8.600E -06
	IKE1	-	1.762E -06	5.920E -06	7.196E -06	8.162E -06
	JAE	-	1.797E -06	6.419E -06	8.008E -06	9.292E -06
	PSI1	-	1.692E -06	5.112E -06	6.013E -06	6.639E -06
	STU	-	1.736E -06	5.518E -06	6.614E -06	7.410E -06
	Average	-	1.633E -06	5.363E -06	6.521E -06	7.410E -06
Eu-153	BEN	-	1.064E -06	4.492E -06	6.042E -06	7.438E -06
	BNFL	-	1.064E -06	4.492E -06	6.042E -06	7.439E -06
	CEA	-	1.195E -06	5.409E -06	7.223E -06	8.776E -06
	ECN	-	7.594E -07	3.496E -06	4.692E -06	5.722E -06
	EDF	-	-	-	-	-
	HIT	-	1.117E -06	4.800E -06	6.396E -06	7.790E -06
	IKE1	-	1.114E -06	4.965E -06	6.697E -06	8.227E -06
	JAE	-	1.106E -06	4.642E -06	6.178E -06	7.529E -06
	PSI1	-	1.199E -06	5.428E -06	7.196E -06	8.659E -06
	STU	-	1.190E -06	5.312E -06	7.117E -06	8.690E -06
	Average	-	1.090E -06	4.782E -06	6.398E -06	7.808E -06

Tab. 16: Nuclide Densities of Fiss. Prod:, Benchmark A,(cont.)

Nuclide	Contributor	Burnup, MWd/kg				
		0.0	10.0	33.0	42.0	50.0
Eu-154	BEN	-	1.156E -07	1.456E -06	2.433E -06	3.503E -06
	BNFL	-	1.156E -07	1.456E -06	2.433E -06	3.504E -06
	CEA	-	1.242E -07	1.440E -06	2.284E -06	3.123E -06
	ECN	-	7.654E -08	9.051E -07	1.447E -06	1.991E -06
	EDF	-	-	-	-	-
	HIT	-	1.122E -07	1.184E -06	1.836E -06	2.473E -06
	IKE1	-	1.109E -07	1.252E -06	2.002E -06	2.770E -06
	JAE	-	1.092E -07	1.147E -06	1.785E -06	2.413E -06
	PSI1	-	1.203E -07	1.434E -06	2.279E -06	3.116E -06
	STU	-	1.247E -07	1.413E -06	2.244E -06	3.087E -06
	Average	-	1.121E -07	1.299E -06	2.083E -06	2.886E -06
Eu-155	BEN	-	2.335E -07	3.816E -07	4.583E -07	5.492E -07
	BNFL	-	2.335E -07	3.818E -07	4.584E -07	5.494E -07
	CEA	-	3.071E -07	7.462E -07	9.758E -07	1.212E -06
	ECN	-	1.784E -07	4.335E -07	5.774E -07	7.305E -07
	EDF	-	-	-	-	-
	HIT	-	1.662E -07	3.156E -07	4.162E -07	5.224E -07
	IKE1	-	2.879E -07	6.855E -07	8.893E -07	1.105E -06
	JAE	-	1.665E -07	3.064E -07	3.959E -07	4.897E -07
	PSI1	-	3.804E -07	8.609E -07	1.092E -06	1.329E -06
	STU	-	3.959E -07	7.413E -08	1.107E -06	1.345E -06
	Average	-	2.610E -07	4.651E -07	7.077E -07	8.702E -07

Tab. 17: Fractional Absorption Rates of Actinides, Benchmark B

Nuclide	Contributor	Burnup, MWd/kg				
		0.0	10.0	33.0	42.0	50.0
U-234	BNFL	5.262E -05	4.781E -05	3.614E -05	3.263E -05	2.919E -05
	CEA	-	6.710E -05	8.747E -05	9.620E -05	1.050E -04
	ECN	5.767E -05	7.427E -05	1.013E -04	1.102E -04	1.179E -04
	EDF	5.480E -05	7.030E -05	9.176E -05	9.601E -05	9.823E -05
	HIT	5.195E -05	6.605E -05	8.909E -05	9.574E -05	1.031E -04
	IKE1	5.299E -05	7.004E -05	9.806E -05	1.071E -04	1.148E -04
	IKE2	5.850E -05	-	-	-	-
	JAE	5.571E -05	7.068E -05	9.482E -05	1.030E -04	1.102E -04
	PSI1	5.833E -05	7.326E -05	9.669E -05	1.041E -04	1.106E -04
	STU	5.466E -05	7.219E -05	1.019E -04	1.116E -04	1.199E -04
	Average	5.525E -05	6.797E -05	8.859E -05	9.517E -05	1.010E -04
U-235	BNFL	1.454E -02	1.300E -02	9.767E -03	8.670E -03	7.542E -03
	CEA	1.445E -02	1.249E -02	9.295E -03	8.105E -03	6.261E -03
	ECN	1.442E -02	1.255E -02	9.376E -03	8.185E -03	7.169E -03
	EDF	1.455E -02	1.257E -02	9.246E -03	8.010E -03	6.948E -03
	HIT	1.487E -02	1.287E -02	9.428E -03	8.154E -03	7.062E -03
	IKE1	1.453E -02	1.260E -02	9.359E -03	8.154E -03	7.117E -03
	IKE2	1.441E -02	-	-	-	-
	JAE	1.474E -02	1.272E -02	9.309E -03	8.055E -03	6.989E -03
	PSI1	1.461E -02	1.263E -02	9.334E -03	8.116E -03	7.081E -03
	STU	1.435E -02	1.249E -02	9.356E -03	8.197E -03	7.173E -03
	Average	1.455E -02	1.266E -02	9.386E -03	8.183E -03	7.038E -03
U-236	BNFL	-	1.017E -04	4.148E -04	4.900E -04	5.628E -04
	CEA	-	1.746E -04	4.573E -04	5.317E -04	5.831E -04
	ECN	-	1.773E -04	4.673E -04	5.447E -04	5.988E -04
	EDF	-	1.800E -04	4.676E -04	5.421E -04	5.928E -04
	HIT	-	1.795E -04	4.573E -04	5.234E -04	5.764E -04
	IKE1	-	1.936E -04	5.190E -04	6.068E -04	6.674E -04
	IKE2	-	-	-	-	-
	JAE	-	1.986E -04	5.166E -04	5.989E -04	6.547E -04
	PSI1	-	1.830E -04	4.709E -04	5.474E -04	6.008E -04
	STU	-	1.799E -04	4.855E -04	5.686E -04	6.268E -04
	Average	-	1.742E -04	4.729E -04	5.504E -04	6.070E -04

Tab. 17: Fractional Absorption Rates of Actinides, Benchmark B, (cont.)

Nuclide	Contributor	Burnup, MWd/kg				
		0.0	10.0	33.0	42.0	50.0
U-238	BNFL	2.360E -01	2.306E -01	2.325E -01	2.322E -01	2.339E -01
	CEA	2.347E -01	2.333E -01	2.338E -01	2.344E -01	2.350E -01
	ECN	2.392E -01	2.387E -01	2.404E -01	2.412E -01	2.419E -01
	EDF	2.335E -01	2.328E -01	2.338E -01	2.344E -01	2.350E -01
	HIT	2.370E -01	2.358E -01	2.366E -01	2.372E -01	2.373E -01
	IKE1	2.361E -01	2.345E -01	2.338E -01	2.339E -01	2.341E -01
	IKE2	2.342E -01	-	-	-	-
	JAE	2.359E -01	2.349E -01	2.356E -01	2.360E -01	2.365E -01
	PSI1	2.378E -01	2.369E -01	2.384E -01	2.395E -01	2.405E -01
	STU	2.320E -01	2.305E -01	2.305E -01	2.311E -01	2.317E -01
	Average	2.356E -01	2.342E -01	2.350E -01	2.356E -01	2.362E -01
Np-237	BNFL	-	1.374E -04	7.749E -04	9.762E -04	1.201E -03
	CEA	-	2.194E -04	7.317E -04	9.259E -04	1.091E -03
	ECN	-	2.394E -04	8.090E -04	1.019E -03	1.195E -03
	EDF	-	2.371E -04	7.838E -04	9.886E -04	1.161E -03
	HIT	-	2.658E -04	8.780E -04	1.094E -03	1.279E -03
	IKE1	-	2.028E -04	7.116E -04	9.084E -04	1.077E -03
	IKE2	-	-	-	-	-
	JAE	-	1.906E -04	6.546E -04	8.329E -04	9.842E -04
	PSI1	-	2.668E -04	9.000E -04	1.134E -03	1.330E -03
	STU	-	1.947E -04	6.648E -04	8.473E -04	1.006E -03
	Average	-	2.171E -04	7.676E -04	9.696E -04	1.147E -03
Pu-238	BNFL	3.631E -03	3.361E -03	3.142E -03	3.074E -03	3.037E -03
	CEA	3.738E -03	3.442E -03	3.631E -03	3.967E -03	4.381E -03
	ECN	3.674E -03	3.425E -03	3.768E -03	4.148E -03	4.574E -03
	EDF	3.663E -03	3.397E -03	3.725E -03	4.102E -03	4.520E -03
	HIT	3.826E -03	3.545E -03	3.820E -03	4.174E -03	4.577E -03
	IKE1	3.659E -03	3.398E -03	3.706E -03	4.073E -03	4.480E -03
	IKE2	3.683E -03	-	-	-	-
	JAE	3.645E -03	3.381E -03	3.725E -03	4.122E -03	4.568E -03
	PSI1	3.692E -03	3.423E -03	3.694E -03	4.021E -03	4.387E -03
	STU	3.632E -03	3.385E -03	3.737E -03	4.139E -03	4.578E -03
	Average	3.684E -03	3.417E -03	3.661E -03	3.980E -03	4.345E -03

Tab. 17: Fractional Absorption Rates of Actinides, Benchmark B, (cont.)

Nuclide	Contributor	Burnup, MWd/kg				
		0.0	10.0	33.0	42.0	50.0
	BNFL	4.619E -01	4.131E -01	3.284E -01	3.048E -01	2.853E -01
	CEA	4.605E -01	3.979E -01	3.163E -01	2.928E -01	2.758E -01
	ECN	4.594E -01	4.021E -01	3.234E -01	3.010E -01	2.849E -01
	EDF	4.631E -01	4.012E -01	3.153E -01	2.910E -01	2.737E -01
	HIT	4.637E -01	4.046E -01	3.214E -01	2.975E -01	2.803E -01
Pu-239	IKE1	4.601E -01	4.004E -01	3.174E -01	2.930E -01	2.754E -01
	IKE2	4.598E -01	-	-	-	-
	JAE	4.598E -01	3.999E -01	3.186E -01	2.955E -01	2.789E -01
	PSI1	4.573E -01	3.994E -01	3.214E -01	2.996E -01	2.841E -01
	STU	4.605E -01	4.001E -01	3.163E -01	2.924E -01	2.753E -01
	Average	4.606E -01	4.021E -01	3.198E -01	2.964E -01	2.793E -01
	BNFL	1.310E -01	1.297E -01	1.333E -01	1.321E -01	1.307E -01
	CEA	1.367E -01	1.370E -01	1.349E -01	1.324E -01	1.294E -01
	ECN	1.360E -01	1.370E -01	1.363E -01	1.340E -01	1.313E -01
	EDF	1.399E -01	1.399E -01	1.369E -01	1.336E -01	1.299E -01
	HIT	1.348E -01	1.363E -01	1.357E -01	1.334E -01	1.303E -01
Pu-240	IKE1	1.355E -01	1.361E -01	1.347E -01	1.321E -01	1.289E -01
	IKE2	1.380E -01	-	-	-	-
	JAE	1.386E -01	1.388E -01	1.362E -01	1.333E -01	1.299E -01
	PSI1	1.350E -01	1.356E -01	1.346E -01	1.325E -01	1.300E -01
	STU	1.409E -01	1.410E -01	1.382E -01	1.353E -01	1.316E -01
	Average	1.366E -01	1.368E -01	1.357E -01	1.332E -01	1.302E -01
	BNFL	9.875E -02	1.020E -01	1.280E -01	1.326E -01	1.367E -01
	CEA	9.844E -02	1.107E -01	1.328E -01	1.363E -01	1.369E -01
	ECN	9.809E -02	1.096E -01	1.329E -01	1.370E -01	1.384E -01
	EDF	9.601E -02	1.085E -01	1.327E -01	1.367E -01	1.375E -01
	HIT	9.695E -02	1.084E -01	1.320E -01	1.361E -01	1.372E -01
Pu-241	IKE1	9.814E -02	1.091E -01	1.312E -01	1.350E -01	1.360E -01
	IKE2	9.780E -02	-	-	-	-
	JAE	9.813E -02	1.100E -01	1.330E -01	1.366E -01	1.374E -01
	PSI1	9.849E -02	1.085E -01	1.294E -01	1.331E -01	1.344E -01
	STU	9.793E -02	1.106E -01	1.352E -01	1.396E -01	1.407E -01
	Average	9.787E -02	1.086E -01	1.319E -01	1.359E -01	1.373E -01

Tab. 17: Fractional Absorption Rates of Actinides, Benchmark B, (cont.)

Nuclide	Contributor	Burnup, MWd/kg				
		0.0	10.0	33.0	42.0	50.0
	BNFL	1.573E -02	1.621E -02	2.039E -02	2.185E -02	2.361E -02
	CEA	1.260E -02	1.300E -02	1.738E -02	1.878E -02	1.998E -02
	ECN	1.258E -02	1.375E -02	1.722E -02	1.864E -02	1.985E -02
	EDF	1.292E -02	1.416E -02	1.798E -02	1.956E -02	2.089E -02
	HIT	1.049E -02	1.167E -02	1.520E -02	1.679E -02	1.827E -02
Pu-242	IKE1	1.215E -02	1.347E -02	1.695E -02	1.859E -02	2.001E -02
	IKE2	1.274E -02	-	-	-	-
	JAE	1.226E -02	1.352E -02	1.746E -02	1.907E -02	2.044E -02
	PSI1	1.207E -02	1.325E -02	1.668E -02	1.809E -02	1.930E -02
	STU	1.114E -02	1.224E -02	1.563E -02	1.692E -02	1.804E -02
	Average	1.247E -02	1.347E -02	1.721E -02	1.870E -02	2.004E -02
	BNFL	-	1.733E -03	6.562E -03	7.382E -03	7.926E -03
	CEA	-	2.203E -03	6.189E -03	7.216E -03	7.833E -03
	ECN	-	2.919E -03	7.182E -03	7.941E -03	8.219E -03
	EDF	-	2.907E -03	7.211E -03	7.967E -03	8.222E -03
	HIT	-	2.714E -03	6.920E -03	7.735E -03	8.067E -03
Am-241	IKE1	-	2.887E -03	7.096E -03	7.826E -03	8.076E -03
	IKE2	-	-	-	-	-
	JAE	-	2.783E -03	7.069E -03	7.874E -03	8.189E -03
	PSI1	-	2.688E -03	6.647E -03	7.364E -03	7.654E -03
	STU	-	2.935E -03	7.302E -03	8.073E -03	8.306E -03
	Average	-	2.641E -03	6.909E -03	7.709E -03	8.055E -03
	BNFL	-	7.802E -05	7.193E -04	8.553E -04	9.533E -04
	CEA	-	1.386E -04	6.503E -04	7.947E -04	8.842E -04
	ECN	-	1.622E -04	6.710E -04	7.784E -04	8.245E -04
	EDF	-	1.767E -04	7.678E -04	8.939E -04	9.482E -04
	HIT	-	-	-	-	-
Am-242m	IKE1	-	1.580E -04	6.660E -04	7.719E -04	8.167E -04
	IKE2	-	-	-	-	-
	JAE	-	2.340E -04	1.044E -03	1.229E -03	1.315E -03
	PSI1	-	-	-	-	-
	STU	-	1.775E -04	7.527E -04	8.723E -04	9.250E -04
	Average	-	1.607E -04	7.531E -04	8.851E -04	9.524E -04

Tab. 17: Fractional Absorption Rates of Actinides, Benchmark B, (cont.)

Nuclide	Contributor	Burnup, MWd/kg				
		0.0	10.0	33.0	42.0	50.0
Am-243	BNFL	-	2.539E -03	1.059E -02	1.285E -02	1.536E -02
	CEA	-	3.344E -03	9.302E -03	1.134E -02	1.306E -02
	ECN	-	3.343E -03	9.384E -03	1.144E -02	1.318E -02
	EDF	-	3.438E -03	9.633E -03	1.178E -02	1.361E -02
	HIT	-	2.685E -03	7.804E -03	9.664E -03	1.126E -02
	IKE1	-	3.193E -03	9.026E -03	1.107E -02	1.279E -02
	IKE2	-	-	-	-	-
	JAE	-	3.194E -03	9.122E -03	1.123E -02	1.304E -02
	PSI1	-	2.998E -03	8.844E -03	1.092E -02	1.265E -02
	STU	-	2.975E -03	8.477E -03	1.038E -02	1.203E -02
	Average	-	3.079E -03	9.131E -03	1.118E -02	1.300E -02
Cm-242	BNFL	-	-	-	-	-
	CEA	-	2.275E -05	1.336E -04	1.695E -04	1.925E -04
	ECN	-	2.533E -05	1.347E -04	1.699E -04	1.936E -04
	EDF	-	3.825E -05	2.096E -04	2.682E -04	3.091E -04
	HIT	-	2.376E -05	1.273E -04	1.625E -04	1.870E -04
	IKE1	-	2.506E -05	1.337E -04	1.691E -04	1.930E -04
	IKE2	-	-	-	-	-
	JAE	-	2.731E -05	1.494E -04	1.901E -04	2.179E -04
	PSI1	-	2.319E -05	1.226E -04	1.547E -04	1.753E -04
	STU	-	2.578E -05	1.397E -04	1.771E -04	2.025E -04
	Average	-	2.643E -05	1.438E -04	1.826E -04	2.089E -04
Cm-243	BNFL	-	-	-	-	-
	CEA	-	2.985E -06	5.340E -05	8.349E -05	1.095E -04
	ECN	-	2.848E -06	4.892E -05	7.662E -05	1.015E -04
	EDF	-	4.698E -06	8.121E -05	1.282E -04	1.708E -04
	HIT	-	-	-	-	-
	IKE1	-	3.021E -06	5.061E -05	7.909E -05	1.044E -04
	IKE2	-	-	-	-	-
	JAE	-	2.588E -06	4.750E -05	7.595E -05	1.021E -04
	PSI1	-	2.814E -06	4.652E -05	7.242E -05	9.516E -05
	STU	-	2.873E -06	5.016E -05	7.902E -05	1.054E -04
	Average	-	3.118E -06	5.404E -05	8.497E -05	1.127E -04

Tab. 17: Fractional Absorption Rates of Actinides, Benchmark B, (cont.)

Nuclide	Contributor	Burnup, MWd/kg				
		0.0	10.0	33.0	42.0	50.0
Cm-244	BNFL	-	-	-	-	-
	CEA	-	1.850E -04	1.666E -03	2.533E -03	3.404E -03
	ECN	-	-	-	-	-
	EDF	-	-	-	-	-
	HIT	-	1.352E -04	1.146E -03	1.749E -03	2.363E -03
	IKE1	-	1.445E -04	1.352E -03	2.090E -03	2.851E -03
	IKE2	-	-	-	-	-
	JAE	-	1.774E -04	1.636E -03	2.490E -03	3.339E -03
	PSI1	-	-	-	-	-
	STU	-	1.604E -04	1.475E -03	2.264E -03	3.085E -03
	Average	-	1.605E -04	1.455E -03	2.225E -03	3.009E -03
Cm-245	BNFL	-	-	-	-	-
	CEA	-	-	-	-	-
	ECN	-	-	-	-	-
	EDF	-	-	-	-	-
	HIT	-	1.361E -05	4.375E -04	8.218E -04	1.285E -03
	IKE1	-	1.732E -05	5.190E -04	9.996E -04	1.577E -03
	IKE2	-	-	-	-	-
	JAE	-	1.979E -05	6.077E -04	1.163E -03	1.824E -03
	PSI1	-	-	-	-	-
	STU	-	1.886E -05	5.685E -04	1.090E -03	1.730E -03
	Average	-	1.739E -05	5.332E -04	1.019E -03	1.604E -03

Tab. 18: Fract. Absorption Rates of Fiss. Prod., Benchmark B

Nuclide	Contributor	Burnup, MWd/kg 0.0	10.0	33.0	42.0	50.0
Mo-95	BNFL	-	2.198E -04	1.090E -03	1.374E -03	1.705E -03
	CEA	-	1.809E -04	1.217E -03	1.628E -03	1.986E -03
	ECN	-	2.374E -04	1.246E -03	1.623E -03	1.947E -03
	EDF	-	-	-	-	-
	HIT	-	2.318E -04	1.178E -03	1.534E -03	1.837E -03
	IKE1	-	2.218E -04	1.218E -03	1.607E -03	1.948E -03
	JAE	-	4.491E -04	1.309E -03	1.609E -03	1.865E -03
	PSI1	-	2.235E -04	1.242E -03	1.638E -03	1.985E -03
	STU	-	-	-	-	-
	Average	-	2.520E -04	1.214E -03	1.573E -03	1.896E -03
Tc-99	BNFL	-	4.451E -04	2.140E -03	2.678E -03	3.295E -03
	CEA	-	1.155E -03	3.502E -03	4.326E -03	5.025E -03
	ECN	-	1.052E -03	3.102E -03	3.797E -03	4.382E -03
	EDF	-	-	-	-	-
	HIT	-	1.165E -03	3.487E -03	4.258E -03	4.969E -03
	IKE1	-	1.162E -03	3.608E -03	4.484E -03	5.226E -03
	JAE	-	1.158E -03	3.444E -03	4.234E -03	4.899E -03
	PSI1	-	1.265E -03	3.732E -03	4.581E -03	5.296E -03
	STU	-	-	-	-	-
	Average	-	1.057E -03	3.288E -03	4.051E -03	4.727E -03
Ru-101	BNFL	-	-	-	-	-
	CEA	-	5.443E -04	1.764E -03	2.229E -03	2.635E -03
	ECN	-	-	-	-	-
	EDF	-	-	-	-	-
	HIT	-	4.912E -04	1.564E -03	1.968E -03	2.327E -03
	IKE1	-	4.802E -04	1.565E -03	1.980E -03	2.344E -03
	JAE	-	4.925E -04	1.576E -03	1.981E -03	2.334E -03
	PSI1	-	-	-	-	-
	STU	-	-	-	-	-
	Average	-	5.021E -04	1.617E -03	2.040E -03	2.410E -03

Tab. 18: Fract. Absorption Rates of Fiss. Prod., Benchmark B,(cont.)

Nuclide	Contributor	Burnup, MWd/kg 0.0	10.0	33.0	42.0	50.0
Rh-103	BNFL	-	-	-	-	-
	CEA	-	2.593E -03	9.028E -03	1.101E -02	1.257E -02
	ECN	-	-	-	-	-
	EDF	-	-	-	-	-
	HIT	-	2.767E -03	8.987E -03	1.093E -02	1.235E -02
	IKE1	-	2.674E -03	8.645E -03	1.047E -02	1.192E -02
	JAE	-	2.756E -03	8.945E -03	1.084E -02	1.234E -02
	PSI1	-	-	-	-	-
	STU	-	3.590E -03	9.933E -03	1.194E -02	1.351E -02
	Average	-	2.876E -03	9.108E -03	1.104E -02	1.254E -02
Pd-105	BNFL	-	-	-	-	-
	CEA	-	4.711E -04	1.594E -03	2.044E -03	2.449E -03
	ECN	-	-	-	-	-
	EDF	-	-	-	-	-
	HIT	-	4.681E -04	1.550E -03	1.978E -03	2.354E -03
	IKE1	-	4.523E -04	1.521E -03	1.947E -03	2.329E -03
	JAE	-	4.633E -04	1.544E -03	1.968E -03	2.346E -03
	PSI1	-	-	-	-	-
	STU	-	-	-	-	-
	Average	-	4.637E -04	1.552E -03	1.984E -03	2.369E -03
Pd-107	BNFL	-	-	-	-	-
	CEA	-	3.116E -04	1.029E -03	1.306E -03	1.550E -03
	ECN	-	-	-	-	-
	EDF	-	-	-	-	-
	HIT	-	3.083E -04	1.004E -03	1.271E -03	1.503E -03
	IKE1	-	3.491E -04	1.193E -03	1.538E -03	1.852E -03
	JAE	-	3.238E -04	1.055E -03	1.334E -03	1.577E -03
	PSI1	-	-	-	-	-
	STU	-	-	-	-	-
	Average	-	3.232E -04	1.070E -03	1.362E -03	1.620E -03

Tab. 18: Fract. Absorption Rates of Fiss. Prod., Benchmark B,(cont.)

Nuclide	Contributor	Burnup, MWd/kg				
		0.0	10.0	33.0	42.0	50.0
Pd-108	BNFL	-	-	-	-	-
	CEA	-	3.198E -04	1.086E -03	1.395E -03	1.672E -03
	ECN	-	-	-	-	-
	EDF	-	-	-	-	-
	HIT	-	3.811E -04	1.183E -03	1.495E -03	1.769E -03
	IKE1	-	3.341E -04	1.145E -03	1.478E -03	1.780E -03
	JAE	-	4.216E -04	1.253E -03	1.547E -03	1.798E -03
	PSI1	-	-	-	-	-
	STU	-	-	-	-	-
	Average	-	3.642E -04	1.167E -03	1.479E -03	1.755E -03
Ag-109	BNFL	-	6.305E -04	2.718E -03	3.289E -03	3.898E -03
	CEA	-	1.436E -03	3.723E -03	4.392E -03	4.915E -03
	ECN	-	8.912E -04	2.171E -03	2.496E -03	2.733E -03
	EDF	-	-	-	-	-
	HIT	-	1.301E -03	3.398E -03	3.999E -03	4.538E -03
	IKE1	-	1.491E -03	3.995E -03	4.730E -03	5.297E -03
	JAE	-	1.558E -03	3.999E -03	4.691E -03	5.223E -03
	PSI1	-	1.401E -03	3.694E -03	4.404E -03	4.977E -03
	STU	-	1.352E -03	3.443E -03	3.978E -03	4.334E -03
	Average	-	1.258E -03	3.393E -03	3.997E -03	4.489E -03
Xe-131	BNFL	-	1.065E -03	4.576E -03	5.480E -03	6.396E -03
	CEA	-	2.060E -03	5.817E -03	6.870E -03	7.638E -03
	ECN	-	1.774E -03	4.700E -03	5.497E -03	6.083E -03
	EDF	-	-	-	-	-
	HIT	-	1.740E -03	4.792E -03	5.673E -03	6.369E -03
	IKE1	-	2.085E -03	5.832E -03	6.857E -03	7.589E -03
	JAE	-	1.720E -03	4.160E -03	4.866E -03	5.424E -03
	PSI1	-	1.962E -03	5.303E -03	6.277E -03	7.018E -03
	STU	-	2.217E -03	6.044E -03	7.116E -03	7.880E -03
	Average	-	1.828E -03	5.153E -03	6.079E -03	6.800E -03

Tab. 18: Fract. Absorption Rates of Fiss. Prod., Benchmark B,(cont.)

Nuclide	Contributor	Burnup, MWd/kg				
		0.0	10.0	33.0	42.0	50.0
Xe-135	BNFL	-	-	-	-	-
	CEA	-	1.459E -02	1.387E -02	1.366E -02	1.347E -02
	ECN	-	-	-	-	-
	EDF	-	-	-	-	-
	HIT	-	1.290E -02	1.368E -02	1.389E -02	1.409E -02
	IKE1	-	1.294E -02	1.368E -02	1.391E -02	1.408E -02
	JAE	-	1.295E -02	1.368E -02	1.395E -02	1.412E -02
	PSI1	-	-	-	-	-
	STU	-	1.254E -02	1.350E -02	1.374E -02	1.415E -02
	Average	-	1.318E -02	1.368E -02	1.383E -02	1.398E -02
Cs-133	BNFL	-	9.052E -04	4.187E -03	5.168E -03	6.258E -03
	CEA	-	1.816E -03	5.413E -03	6.617E -03	7.606E -03
	ECN	-	1.663E -03	4.875E -03	5.938E -03	6.788E -03
	EDF	-	-	-	-	-
	HIT	-	1.551E -03	4.721E -03	5.773E -03	6.667E -03
	IKE1	-	1.813E -03	5.522E -03	6.789E -03	7.832E -03
	JAE	-	1.573E -03	4.541E -03	5.527E -03	6.346E -03
	PSI1	-	1.675E -03	5.009E -03	6.135E -03	7.069E -03
	STU	-	1.934E -03	5.754E -03	7.055E -03	8.120E -03
	Average	-	1.616E -03	5.003E -03	6.125E -03	7.086E -03
Cs-135	BNFL	-	-	-	-	-
	CEA	-	2.054E -04	6.697E -04	8.475E -04	1.004E -03
	ECN	-	-	-	-	-
	EDF	-	-	-	-	-
	HIT	-	1.984E -04	6.045E -04	7.465E -04	8.633E -04
	IKE1	-	2.410E -04	7.511E -04	9.328E -04	1.087E -03
	JAE	-	2.257E -04	6.808E -04	8.380E -04	9.698E -04
	PSI1	-	-	-	-	-
	STU	-	2.251E -04	6.984E -04	8.645E -04	1.003E -03
	Average	-	2.191E -04	6.809E -04	8.458E -04	9.854E -04

Tab. 18: Fract. Absorption Rates of Fiss. Prod., Benchmark B,(cont.)

Nuclide	Contributor	Burnup, MWd/kg 0.0	10.0	33.0	42.0	50.0
Nd-143	BNFL	-	5.021E -04	2.760E -03	3.602E -03	4.641E -03
	CEA	-	1.012E -03	3.826E -03	5.013E -03	6.092E -03
	ECN	-	1.125E -03	4.224E -03	5.535E -03	6.728E -03
	EDF	-	-	-	-	-
	HIT	-	1.367E -03	5.301E -03	7.024E -03	8.596E -03
	IKE1	-	1.000E -03	3.721E -03	4.861E -03	5.893E -03
	JAE	-	1.032E -03	3.762E -03	4.890E -03	5.909E -03
	PSI1	-	9.954E -04	3.645E -03	4.740E -03	5.728E -03
	STU	-	1.106E -03	3.880E -03	5.060E -03	6.123E -03
	Average	-	1.017E -03	3.890E -03	5.091E -03	6.214E -03
Nd-145	BNFL	-	3.004E -04	1.456E -03	1.826E -03	2.252E -03
	CEA	-	5.763E -04	1.832E -03	2.304E -03	2.717E -03
	ECN	-	6.611E -04	2.154E -03	2.729E -03	3.234E -03
	EDF	-	-	-	-	-
	HIT	-	4.957E -04	1.573E -03	1.980E -03	2.338E -03
	IKE1	-	5.686E -04	1.819E -03	2.288E -03	2.697E -03
	JAE	-	5.107E -04	1.585E -03	1.981E -03	2.326E -03
	PSI1	-	5.497E -04	1.736E -03	2.182E -03	2.574E -03
	STU	-	5.612E -04	1.781E -03	2.238E -03	2.634E -03
	Average	-	5.280E -04	1.742E -03	2.191E -03	2.596E -03
Pm-147	BNFL	-	8.053E -04	2.147E -03	2.269E -03	2.344E -03
	CEA	-	2.103E -03	4.227E -03	4.446E -03	4.500E -03
	ECN	-	2.239E -03	4.481E -03	4.805E -03	4.978E -03
	EDF	-	-	-	-	-
	HIT	-	1.992E -03	3.952E -03	4.199E -03	4.370E -03
	IKE1	-	2.139E -03	4.300E -03	4.571E -03	4.688E -03
	JAE	-	1.959E -03	3.863E -03	4.146E -03	4.293E -03
	PSI1	-	2.085E -03	4.087E -03	4.369E -03	4.512E -03
	STU	-	2.352E -03	4.481E -03	4.742E -03	4.834E -03
	Average	-	1.959E -03	3.942E -03	4.194E -03	4.315E -03

Tab. 18: Fract. Absorption Rates of Fiss. Prod., Benchmark B,(cont.)

Nuclide	Contributor	Burnup, MWd/kg 0.0	10.0	33.0	42.0	50.0
Pm-148m	BNFL	-	4.724E -04	1.508E -03	1.634E -03	1.731E -03
	CEA	-	5.566E -04	1.277E -03	1.372E -03	1.411E -03
	ECN	-	5.090E -04	1.222E -03	1.382E -03	1.469E -03
	EDF	-	-	-	-	-
	HIT	-	4.638E -04	1.115E -03	1.233E -03	1.348E -03
	IKE1	-	5.183E -04	1.272E -03	1.426E -03	1.516E -03
	JAE	-	5.460E -04	1.289E -03	1.459E -03	1.557E -03
	PSI1	-	5.175E -04	1.265E -03	1.406E -03	1.489E -03
	STU	-	5.732E -04	1.354E -03	1.497E -03	1.610E -03
	Average	-	5.196E -04	1.288E -03	1.426E -03	1.516E -03
Sm-149	BNFL	-	5.652E -03	6.786E -03	6.830E -03	6.850E -03
	CEA	-	5.718E -03	6.364E -03	6.372E -03	6.326E -03
	ECN	-	5.278E -03	5.860E -03	5.975E -03	6.006E -03
	EDF	-	-	-	-	-
	HIT	-	5.757E -03	6.339E -03	6.378E -03	6.471E -03
	IKE1	-	5.733E -03	6.412E -03	6.509E -03	6.540E -03
	JAE	-	5.750E -03	6.425E -03	6.582E -03	6.633E -03
	PSI1	-	5.752E -03	6.469E -03	6.607E -03	6.645E -03
	STU	-	5.799E -03	6.462E -03	6.452E -03	6.563E -03
	Average	-	5.680E -03	6.389E -03	6.463E -03	6.504E -03
Sm-150	BNFL	-	1.383E -04	9.945E -04	1.311E -03	1.687E -03
	CEA	-	3.179E -04	1.280E -03	1.684E -03	2.047E -03
	ECN	-	2.867E -04	1.136E -03	1.494E -03	1.821E -03
	EDF	-	-	-	-	-
	HIT	-	2.653E -04	1.021E -03	1.344E -03	1.646E -03
	IKE1	-	3.095E -04	1.236E -03	1.632E -03	1.992E -03
	JAE	-	3.127E -04	1.186E -03	1.541E -03	1.862E -03
	PSI1	-	1.763E -04	7.524E -04	1.018E -03	1.271E -03
	STU	-	3.681E -04	1.433E -03	1.871E -03	2.259E -03
	Average	-	2.718E -04	1.130E -03	1.487E -03	1.823E -03

Tab. 18: Fract. Absorption Rates of Fiss. Prod., Benchmark B,(cont.)

Nuclide	Contributor	Burnup, MWd/kg				
		0.0	10.0	33.0	42.0	50.0
Sm-151	BNFL	-	1.297E -03	3.663E -03	4.025E -03	4.408E -03
	CEA	-	2.646E -03	4.028E -03	4.374E -03	4.680E -03
	ECN	-	2.153E -03	3.198E -03	3.465E -03	3.722E -03
	EDF	-	-	-	-	-
	HIT	-	2.686E -03	3.890E -03	4.125E -03	4.353E -03
	IKE1	-	2.654E -03	4.047E -03	4.381E -03	4.679E -03
	JAE	-	2.656E -03	3.990E -03	4.287E -03	4.542E -03
	PSI1	-	2.515E -03	3.461E -03	3.642E -03	3.826E -03
	STU	-	2.630E -03	4.118E -03	4.490E -03	4.838E -03
	Average	-	2.404E -03	3.799E -03	4.099E -03	4.381E -03
Sm-152	BNFL	-	6.500E -04	3.468E -03	4.202E -03	4.921E -03
	CEA	-	1.590E -03	4.562E -03	5.252E -03	5.735E -03
	ECN	-	1.277E -03	3.531E -03	4.013E -03	4.354E -03
	EDF	-	-	-	-	-
	HIT	-	1.340E -03	3.842E -03	4.527E -03	5.002E -03
	IKE1	-	1.353E -03	4.184E -03	4.918E -03	5.457E -03
	JAE	-	1.187E -03	3.540E -03	4.206E -03	4.718E -03
	PSI1	-	1.624E -03	4.325E -03	4.886E -03	5.260E -03
	STU	-	1.577E -03	4.579E -03	5.330E -03	5.848E -03
	Average	-	1.325E -03	4.004E -03	4.667E -03	5.162E -03
Eu-153	BNFL	-	2.857E -04	2.140E -03	2.913E -03	3.823E -03
	CEA	-	5.920E -04	2.873E -03	3.840E -03	4.653E -03
	ECN	-	4.518E -04	2.158E -03	2.857E -03	3.431E -03
	EDF	-	-	-	-	-
	HIT	-	5.713E -04	2.614E -03	3.483E -03	4.237E -03
	IKE1	-	5.421E -04	2.589E -03	3.496E -03	4.278E -03
	JAE	-	5.354E -04	2.391E -03	3.196E -03	3.896E -03
	PSI1	-	6.153E -04	2.905E -03	3.819E -03	4.550E -03
	STU	-	6.151E -04	2.919E -03	3.906E -03	4.748E -03
	Average	-	5.261E -04	2.574E -03	3.439E -03	4.202E -03

Tab. 18: Fract. Absorption Rates of Fiss. Prod., Benchmark B,(cont.)

Nuclide	Contributor	Burnup, MWd/kg				
		0.0	10.0	33.0	42.0	50.0
Eu-154	BNFL	-	1.948E -05	6.785E -04	1.171E -03	1.911E -03
	CEA	-	1.067E -04	1.335E -03	2.136E -03	2.926E -03
	ECN	-	8.120E -05	1.004E -03	1.597E -03	2.172E -03
	EDF	-	-	-	-	-
	HIT	-	1.204E -04	1.600E -03	2.477E -03	3.310E -03
	IKE1	-	9.783E -05	1.180E -03	1.903E -03	2.631E -03
	JAE	-	1.238E -04	1.385E -03	2.168E -03	2.925E -03
	PSI1	-	1.061E -04	1.327E -03	2.103E -03	2.850E -03
	STU	-	1.131E -04	1.366E -03	2.183E -03	3.004E -03
	Average	-	9.609E -05	1.234E -03	1.967E -03	2.716E -03
Eu-155	BNFL	-	5.174E -04	1.192E -03	1.630E -03	2.329E -03
	CEA	-	5.780E -04	1.657E -03	2.398E -03	3.169E -03
	ECN	-	4.061E -04	1.158E -03	1.698E -03	2.261E -03
	EDF	-	-	-	-	-
	HIT	-	6.892E -04	1.986E -03	2.825E -03	3.605E -03
	IKE1	-	5.425E -04	1.500E -03	2.165E -03	2.873E -03
	JAE	-	7.102E -04	1.795E -03	2.519E -03	3.252E -03
	PSI1	-	7.047E -04	1.821E -03	2.547E -03	3.290E -03
	STU	-	7.352E -04	1.883E -03	2.636E -03	3.450E -03
	Average	-	6.104E -04	1.624E -03	2.302E -03	3.029E -03

Tab. 19: Fractional Fission Rate of Actinides, Benchmark B

Nuclide	Contributor	Burnup, MWd/kg				
		0.0	10.0	33.0	42.0	50.0
U-234	BNFL	1.545E -06	1.415E -06	1.057E -06	9.452E -07	8.352E -07
	CEA	-	2.071E -06	2.817E -06	3.108E -06	3.389E -06
	ECN	1.659E -06	2.213E -06	3.092E -06	3.358E -06	3.581E -06
	EDF	1.630E -06	2.178E -06	2.952E -06	3.101E -06	3.175E -06
	HIT	1.606E -06	2.150E -06	3.035E -06	3.305E -06	3.524E -06
	IKE1	1.629E -06	2.182E -06	3.062E -06	3.328E -06	3.543E -06
	IKE2	1.669E -06	-	-	-	-
	JAE	1.695E -06	2.260E -06	3.178E -06	3.469E -06	3.719E -06
	PSI1	1.642E -06	2.156E -06	2.964E -06	3.209E -06	3.414E -06
	STU	1.903E -06	2.549E -06	3.596E -06	3.913E -06	4.173E -06
	Average	1.664E -06	2.130E -06	2.861E -06	3.082E -06	3.261E -06
U-235	BNFL	1.111E -02	9.928E -03	7.556E -03	6.739E -03	5.893E -03
	CEA	1.116E -02	9.657E -03	7.271E -03	6.371E -03	5.584E -03
	ECN	1.113E -02	9.705E -03	7.331E -03	6.430E -03	5.651E -03
	EDF	1.115E -02	9.648E -03	7.181E -03	6.250E -03	5.444E -03
	HIT	1.138E -02	9.856E -03	7.305E -03	6.346E -03	5.516E -03
	IKE1	1.112E -02	9.655E -03	7.257E -03	6.353E -03	5.568E -03
	IKE2	1.114E -02	-	-	-	-
	JAE	1.108E -02	9.570E -03	7.096E -03	6.172E -03	5.379E -03
	PSI1	1.119E -02	9.686E -03	7.237E -03	6.319E -03	5.534E -03
	STU	1.106E -02	9.643E -03	7.313E -03	6.439E -03	5.658E -03
	Average	1.115E -02	9.705E -03	7.283E -03	6.380E -03	5.581E -03
U-236	BNFL	-	2.827E -06	1.196E -05	1.417E -05	1.628E -05
	CEA	-	6.855E -06	1.869E -05	2.195E -05	2.424E -05
	ECN	-	6.827E -06	1.869E -05	2.197E -05	2.428E -05
	EDF	-	6.808E -06	1.851E -05	2.171E -05	2.393E -05
	HIT	-	6.962E -06	1.885E -05	2.201E -05	2.426E -05
	IKE1	-	7.027E -06	1.924E -05	2.261E -05	2.492E -05
	IKE2	-	-	-	-	-
	JAE	-	7.734E -06	2.081E -05	2.430E -05	2.669E -05
	PSI1	-	6.902E -06	1.871E -05	2.199E -05	2.431E -05
	STU	-	7.762E -06	2.139E -05	2.516E -05	2.780E -05
	Average	-	6.634E -06	1.854E -05	2.176E -05	2.408E -05

Tab. 19: Fractional Fission Rate of Actinides, Benchmark B, (cont.)

Nuclide	Contributor	Burnup, MWd/kg				
		0.0	10.0	33.0	42.0	50.0
U-238	BNFL	2.872E -02	2.812E -02	2.803E -02	2.783E -02	2.782E -02
	CEA	2.908E -02	2.890E -02	2.847E -02	2.830E -02	2.815E -02
	ECN	2.823E -02	2.807E -02	2.767E -02	2.751E -02	2.736E -02
	EDF	2.750E -02	2.735E -02	2.699E -02	2.685E -02	2.673E -02
	HIT	2.840E -02	2.808E -02	2.773E -02	2.753E -02	2.726E -02
	IKE1	2.834E -02	2.822E -02	2.786E -02	2.776E -02	2.759E -02
	IKE2	2.823E -02	-	-	-	-
	JAE	2.966E -02	2.928E -02	2.864E -02	2.840E -02	2.818E -02
	PSI1	2.878E -02	2.857E -02	2.813E -02	2.799E -02	2.787E -02
	STU	3.302E -02	3.289E -02	3.256E -02	3.242E -02	3.227E -02
	Average	2.900E -02	2.883E -02	2.845E -02	2.829E -02	2.814E -02
Np-237	BNFL	-	3.191E -06	1.741E -05	2.158E -05	2.602E -05
	CEA	-	5.072E -06	1.635E -05	2.031E -05	2.350E -05
	ECN	-	5.399E -06	1.764E -05	2.182E -05	2.516E -05
	EDF	-	5.302E -06	1.704E -05	2.116E -05	2.447E -05
	HIT	-	6.077E -06	1.956E -05	2.399E -05	2.755E -05
	IKE1	-	4.572E -06	1.553E -05	1.950E -05	2.272E -05
	IKE2	-	-	-	-	-
	JAE	-	4.515E -06	1.493E -05	1.864E -05	2.164E -05
	PSI1	-	6.196E -06	2.017E -05	2.497E -05	2.881E -05
	STU	-	5.184E -06	1.711E -05	2.138E -05	2.491E -05
	Average	-	5.057E -06	1.730E -05	2.148E -05	2.498E -05
Pu-238	BNFL	5.179E -04	4.812E -04	4.040E -04	3.799E -04	3.583E -04
	CEA	5.277E -04	4.833E -04	4.585E -04	4.781E -04	5.063E -04
	ECN	5.158E -04	4.755E -04	4.721E -04	4.980E -04	5.284E -04
	EDF	5.141E -04	4.734E -04	4.687E -04	4.942E -04	5.240E -04
	HIT	5.174E -04	4.741E -04	4.625E -04	4.848E -04	5.116E -04
	IKE1	5.135E -04	4.728E -04	4.656E -04	4.900E -04	5.182E -04
	IKE2	5.183E -04	-	-	-	-
	JAE	5.232E -04	4.811E -04	4.799E -04	5.095E -04	5.439E -04
	PSI1	5.153E -04	4.745E -04	4.650E -04	4.867E -04	5.125E -04
	STU	5.448E -04	5.022E -04	4.970E -04	5.242E -04	5.555E -04
	Average	5.208E -04	4.798E -04	4.637E -04	4.828E -04	5.065E -04

Tab. 19: Fractional Fission Rate of Actinides, Benchmark B, (cont.)

Nuclide	Contributor	Burnup, MWd/kg				
		0.0	10.0	33.0	42.0	50.0
Pu-239	BNFL	2.940E -01	2.624E -01	2.082E -01	1.932E -01	1.808E -01
	CEA	2.960E -01	2.551E -01	2.026E -01	1.875E -01	1.767E -01
	ECN	2.954E -01	2.578E -01	2.072E -01	1.928E -01	1.825E -01
	EDF	2.982E -01	2.576E -01	2.021E -01	1.865E -01	1.754E -01
	HIT	2.974E -01	2.587E -01	2.053E -01	1.900E -01	1.791E -01
	IKE1	2.969E -01	2.577E -01	2.040E -01	1.884E -01	1.770E -01
	IKE2	2.955E -01	-	-	-	-
	JAE	2.956E -01	2.564E -01	2.040E -01	1.892E -01	1.785E -01
	PSI1	2.956E -01	2.575E -01	2.069E -01	1.928E -01	1.828E -01
	STU	2.968E -01	2.572E -01	2.029E -01	1.875E -01	1.765E -01
	Average	2.961E -01	2.578E -01	2.048E -01	1.898E -01	1.788E -01
Pu-240	BNFL	1.916E -03	1.943E -03	1.968E -03	1.901E -03	1.809E -03
	CEA	2.230E -03	2.298E -03	2.201E -03	2.091E -03	1.969E -03
	ECN	2.180E -03	2.247E -03	2.153E -03	2.049E -03	1.937E -03
	EDF	1.870E -03	1.920E -03	1.815E -03	1.713E -03	1.604E -03
	HIT	2.135E -03	2.208E -03	2.126E -03	2.021E -03	1.902E -03
	IKE1	2.145E -03	2.210E -03	2.116E -03	2.011E -03	1.893E -03
	IKE2	2.186E -03	-	-	-	-
	JAE	2.133E -03	2.177E -03	2.047E -03	1.932E -03	1.813E -03
	PSI1	2.156E -03	2.220E -03	2.138E -03	2.043E -03	1.938E -03
	STU	2.480E -03	2.541E -03	2.390E -03	2.251E -03	2.102E -03
	Average	2.143E -03	2.196E -03	2.106E -03	2.001E -03	1.885E -03
Pu-241	BNFL	7.448E -02	7.686E -02	9.630E -02	9.975E -02	1.028E -01
	CEA	7.448E -02	8.359E -02	1.001E -01	1.026E -01	1.030E -01
	ECN	7.420E -02	8.285E -02	1.002E -01	1.032E -01	1.042E -01
	EDF	7.148E -02	8.053E -02	9.766E -02	1.003E -01	1.007E -01
	HIT	7.462E -02	8.330E -02	1.009E -01	1.038E -01	1.045E -01
	IKE1	7.387E -02	8.203E -02	9.849E -02	1.013E -01	1.021E -01
	IKE2	7.394E -02	-	-	-	-
	JAE	7.390E -02	8.277E -02	9.987E -02	1.026E -01	1.031E -01
	PSI1	7.419E -02	8.168E -02	9.727E -02	9.999E -02	1.009E -01
	STU	7.441E -02	8.392E -02	1.023E -01	1.055E -01	1.062E -01
	Average	7.396E -02	8.195E -02	9.922E -02	1.021E -01	1.030E -01

Tab. 19: Fractional Fission Rate of Actinides, Benchmark B, (cont.)

Nuclide	Contributor	Burnup, MWd/kg				
		0.0	10.0	33.0	42.0	50.0
Pu-242	BNFL	2.879E -04	2.993E -04	3.894E -04	4.216E -04	4.604E -04
	CEA	2.903E -04	3.320E -04	4.626E -04	5.206E -04	5.723E -04
	ECN	2.836E -04	3.234E -04	4.482E -04	5.038E -04	5.533E -04
	EDF	2.689E -04	3.094E -04	4.432E -04	5.045E -04	5.595E -04
	HIT	2.987E -04	3.418E -04	4.785E -04	5.390E -04	5.908E -04
	IKE1	2.782E -04	3.197E -04	4.465E -04	5.031E -04	5.527E -04
	IKE2	2.851E -04	-	-	-	-
	JAE	2.846E -04	3.281E -04	4.785E -04	5.476E -04	6.093E -04
	PSI1	2.802E -04	3.220E -04	4.484E -04	5.049E -04	5.553E -04
	STU	3.238E -04	3.760E -04	5.413E -04	6.166E -04	6.834E -04
	Average	2.881E -04	3.280E -04	4.596E -04	5.180E -04	5.708E -04
Am-241	BNFL	-	2.873E -05	1.031E -04	1.135E -04	1.188E -04
	CEA	-	3.630E -05	9.688E -05	1.103E -04	1.171E -04
	ECN	-	4.738E -05	1.110E -04	1.199E -04	1.216E -04
	EDF	-	4.433E -05	1.047E -04	1.130E -04	1.142E -04
	HIT	-	4.843E -05	1.163E -04	1.264E -04	1.283E -04
	IKE1	-	4.667E -05	1.093E -04	1.179E -04	1.190E -04
	IKE2	-	-	-	-	-
	JAE	-	4.612E -05	1.101E -04	1.193E -04	1.210E -04
	PSI1	-	4.435E -05	1.043E -04	1.130E -04	1.150E -04
	STU	-	5.248E -05	1.232E -04	1.325E -04	1.330E -04
	Average	-	4.386E -05	1.088E -04	1.184E -04	1.209E -04
Am-242m	BNFL	-	6.497E -05	5.985E -04	7.115E -04	7.929E -04
	CEA	-	1.133E -04	5.304E -04	6.476E -04	7.202E -04
	ECN	-	1.326E -04	5.472E -04	6.345E -04	6.717E -04
	EDF	-	1.506E -04	6.536E -04	7.608E -04	8.069E -04
	HIT	-	-	-	-	-
	IKE1	-	1.315E -04	5.538E -04	6.417E -04	6.789E -04
	IKE2	-	-	-	-	-
	JAE	-	1.964E -04	8.753E -04	1.030E -03	1.101E -03
	PSI1	-	-	-	-	-
	STU	-	1.452E -04	6.142E -04	7.112E -04	7.537E -04
	Average	-	1.335E -04	6.247E -04	7.339E -04	7.894E -04

Tab. 19: Fractional Fission Rate of Actinides, Benchmark B, (cont.)

Nuclide	Contributor	Burnup, MWd/kg				
		0.0	10.0	33.0	42.0	50.0
Am-243	BNFL	-	2.689E -05	1.180E -04	1.444E -04	1.737E -04
	CEA	-	3.647E -05	1.077E -04	1.330E -04	1.545E -04
	ECN	-	3.553E -05	1.051E -04	1.295E -04	1.501E -04
	EDF	-	3.570E -05	1.065E -04	1.322E -04	1.543E -04
	HIT	-	3.988E -05	1.219E -04	1.524E -04	1.788E -04
	IKE1	-	3.399E -05	1.019E -04	1.267E -04	1.475E -04
	IKE2	-	-	-	-	-
	JAE	-	3.714E -05	1.127E -04	1.406E -04	1.645E -04
	PSI1	-	3.391E -05	1.061E -04	1.325E -04	1.547E -04
	STU	-	3.704E -05	1.102E -04	1.360E -04	1.583E -04
	Average	-	3.517E -05	1.100E -04	1.364E -04	1.596E -04
Cm-242	BNFL	-	-	-	-	-
	CEA	-	4.225E -06	2.487E -05	3.161E -05	3.600E -05
	ECN	-	5.159E -06	2.744E -05	3.464E -05	3.951E -05
	EDF	-	7.597E -06	4.017E -05	5.052E -05	5.732E -05
	HIT	-	4.722E -06	2.606E -05	3.328E -05	3.832E -05
	IKE1	-	5.042E -06	2.691E -05	3.410E -05	3.898E -05
	IKE2	-	-	-	-	-
	JAE	-	7.319E -06	3.980E -05	5.051E -05	5.776E -05
	PSI1	-	4.687E -06	2.483E -05	3.138E -05	3.561E -05
	STU	-	5.780E -06	3.129E -05	3.968E -05	4.539E -05
	Average	-	5.566E -06	3.017E -05	3.822E -05	4.361E -05
Cm-243	BNFL	-	-	-	-	-
	CEA	-	2.539E -06	4.534E -05	7.083E -05	9.284E -05
	ECN	-	2.433E -06	4.169E -05	6.525E -05	8.634E -05
	EDF	-	4.196E -06	7.264E -05	1.147E -04	1.530E -04
	HIT	-	-	-	-	-
	IKE1	-	2.595E -06	4.341E -05	6.780E -05	8.946E -05
	IKE2	-	-	-	-	-
	JAE	-	2.265E -06	4.147E -05	6.625E -05	8.902E -05
	PSI1	-	2.416E -06	3.988E -05	6.205E -05	8.149E -05
	STU	-	2.464E -06	4.292E -05	6.755E -05	9.001E -05
	Average	-	2.701E -06	4.676E -05	7.350E -05	9.745E -05

Tab. 19: Fractional Fission Rate of Actinides, Benchmark B, (cont.)

Nuclide	Contributor	Burnup, MWd/kg				
		0.0	10.0	33.0	42.0	50.0
Cm-244	BNFL	-	-	-	-	-
	CEA	-	1.154E -05	1.060E -04	1.625E -04	2.200E -04
	ECN	-	-	-	-	-
	EDF	-	-	-	-	-
	HIT	-	8.840E -06	8.369E -05	1.300E -04	1.784E -04
	IKE1	-	1.038E -05	9.822E -05	1.526E -04	2.086E -04
	IKE2	-	-	-	-	-
	JAE	-	9.454E -06	8.827E -05	1.361E -04	1.854E -04
	PSI1	-	-	-	-	-
	STU	-	1.106E -05	1.038E -04	1.601E -04	2.187E -04
	Average	-	1.026E -05	9.600E -05	1.483E -04	2.022E -04
Cm-245	BNFL	-	-	-	-	-
	CEA	-	-	-	-	-
	ECN	-	-	-	-	-
	EDF	-	-	-	-	-
	HIT	-	1.185E -05	3.794E -04	7.125E -04	1.114E -03
	IKE1	-	1.510E -05	4.525E -04	8.714E -04	1.375E -03
	IKE2	-	-	-	-	-
	JAE	-	1.715E -05	5.263E -04	1.007E -03	1.578E -03
	PSI1	-	-	-	-	-
	STU	-	1.632E -05	4.919E -04	9.428E -04	1.497E -03
	Average	-	1.510E -05	4.625E -04	8.835E -04	1.391E -03

Tab. 20: Neutrons per Fission, Benchmark B

Nuclide	Contributor	Burnup, MWd/kg 0.0	10.0	33.0	42.0	50.0
U-234	BNFL	2.603E+00	2.603E+00	2.604E+00	2.604E+00	2.604E+00
	CEA	-	2.634E+00	2.633E+00	2.632E+00	2.632E+00
	ECN	2.611E+00	2.610E+00	2.609E+00	2.609E+00	2.608E+00
	EDF	2.630E+00	2.630E+00	2.629E+00	2.628E+00	2.628E+00
	HIT	2.644E+00	2.643E+00	2.643E+00	2.643E+00	2.643E+00
	IKE1	2.630E+00	2.630E+00	2.629E+00	2.629E+00	2.628E+00
	IKE2	2.634E+00	-	-	-	-
	JAE	2.632E+00	2.631E+00	2.631E+00	2.631E+00	2.631E+00
	PSI1	2.635E+00	2.635E+00	2.633E+00	2.632E+00	2.632E+00
	STU	2.352E+00	2.352E+00	2.352E+00	2.352E+00	2.352E+00
	Average	2.597E+00	2.596E+00	2.596E+00	2.596E+00	2.595E+00
U-235	BNFL	2.439E+00	2.439E+00	2.438E+00	2.438E+00	2.437E+00
	CEA	2.443E+00	2.443E+00	2.442E+00	2.442E+00	2.442E+00
	ECN	2.450E+00	2.450E+00	2.448E+00	2.447E+00	2.447E+00
	EDF	2.427E+00	2.427E+00	2.426E+00	2.426E+00	2.425E+00
	HIT	2.432E+00	2.431E+00	2.431E+00	2.431E+00	2.431E+00
	IKE1	2.443E+00	2.443E+00	2.442E+00	2.442E+00	2.442E+00
	IKE2	2.443E+00	-	-	-	-
	JAE	2.438E+00	2.438E+00	2.437E+00	2.437E+00	2.437E+00
	PSI1	2.443E+00	2.443E+00	2.442E+00	2.442E+00	2.442E+00
	STU	2.422E+00	2.422E+00	2.422E+00	2.422E+00	2.422E+00
	Average	2.438E+00	2.437E+00	2.437E+00	2.436E+00	2.436E+00
U-236	BNFL	2.779E+00	2.779E+00	2.779E+00	2.779E+00	2.779E+00
	CEA	-	2.569E+00	2.574E+00	2.575E+00	2.576E+00
	ECN	2.549E+00	2.552E+00	2.555E+00	2.555E+00	2.556E+00
	EDF	-	2.565E+00	2.570E+00	2.571E+00	2.572E+00
	HIT	-	2.636E+00	2.643E+00	2.645E+00	2.645E+00
	IKE1	2.557E+00	2.559E+00	2.561E+00	2.561E+00	2.562E+00
	IKE2	-	-	-	-	-
	JAE	2.630E+00	2.626E+00	2.629E+00	2.630E+00	2.630E+00
	PSI1	2.569E+00	2.569E+00	2.574E+00	2.575E+00	2.575E+00
	STU	2.317E+00	2.317E+00	2.317E+00	2.317E+00	2.317E+00
	Average	2.567E+00	2.575E+00	2.578E+00	2.579E+00	2.579E+00

Tab. 20: Neutrons per Fission, Benchmark B, (cont.)

Nuclide	Contributor	Burnup, MWd/kg 0.0	10.0	33.0	42.0	50.0
U-238	BNFL	2.802E+00	2.802E+00	2.802E+00	2.802E+00	2.802E+00
	CEA	2.801E+00	2.798E+00	2.797E+00	2.810E+00	2.799E+00
	ECN	2.737E+00	2.735E+00	2.735E+00	2.735E+00	2.735E+00
	EDF	2.816E+00	2.815E+00	2.814E+00	2.814E+00	2.814E+00
	HIT	2.819E+00	2.818E+00	2.817E+00	2.817E+00	2.818E+00
	IKE1	2.811E+00	2.811E+00	2.811E+00	2.811E+00	2.811E+00
	IKE2	2.810E+00	-	-	-	-
	JAE	2.792E+00	2.791E+00	2.791E+00	2.791E+00	2.791E+00
	PSI1	2.822E+00	2.822E+00	2.821E+00	2.821E+00	2.821E+00
	STU	2.441E+00	2.441E+00	2.441E+00	2.441E+00	2.441E+00
	Average	2.765E+00	2.759E+00	2.759E+00	2.760E+00	2.759E+00
Np-237	BNFL	2.868E+00	2.868E+00	2.868E+00	2.869E+00	2.869E+00
	CEA	-	2.879E+00	2.879E+00	2.879E+00	2.879E+00
	ECN	2.858E+00	2.857E+00	2.857E+00	2.856E+00	2.856E+00
	EDF	3.160E+00	2.875E+00	2.875E+00	2.875E+00	2.875E+00
	HIT	-	2.841E+00	2.840E+00	2.841E+00	2.841E+00
	IKE1	2.875E+00	2.875E+00	2.875E+00	2.875E+00	2.875E+00
	IKE2	-	-	-	-	-
	JAE	2.858E+00	2.857E+00	2.856E+00	2.856E+00	2.855E+00
	PSI1	2.881E+00	2.881E+00	2.880E+00	2.880E+00	2.879E+00
	STU	2.534E+00	2.534E+00	2.534E+00	2.534E+00	2.534E+00
	Average	2.862E+00	2.830E+00	2.829E+00	2.829E+00	2.829E+00
Pu-238	BNFL	3.026E+00	3.026E+00	3.022E+00	3.020E+00	3.018E+00
	CEA	3.030E+00	3.029E+00	3.025E+00	3.023E+00	3.021E+00
	ECN	3.029E+00	3.029E+00	3.024E+00	3.023E+00	3.021E+00
	EDF	2.875E+00	3.026E+00	3.021E+00	3.020E+00	3.018E+00
	HIT	3.032E+00	3.030E+00	3.026E+00	3.024E+00	3.022E+00
	IKE1	3.026E+00	3.026E+00	3.022E+00	3.020E+00	3.018E+00
	IKE2	3.030E+00	-	-	-	-
	JAE	3.031E+00	3.030E+00	3.025E+00	3.023E+00	3.021E+00
	PSI1	3.029E+00	3.028E+00	3.024E+00	3.022E+00	3.020E+00
	STU	2.895E+00	2.895E+00	2.895E+00	2.895E+00	2.895E+00
	Average	3.000E+00	3.013E+00	3.009E+00	3.008E+00	3.006E+00

Tab. 20: Neutrons per Fission, Benchmark B, (cont.)

Nuclide	Contributor	Burnup, MWd/kg				
		0.0	10.0	33.0	42.0	50.0
Pu-239	BNFL	2.876E+00	2.876E+00	2.875E+00	2.875E+00	2.875E+00
	CEA	2.875E+00	2.874E+00	2.873E+00	2.872E+00	2.872E+00
	ECN	2.883E+00	2.882E+00	2.878E+00	2.880E+00	2.878E+00
	EDF	3.026E+00	2.856E+00	2.856E+00	2.856E+00	2.855E+00
	HIT	2.880E+00	2.879E+00	2.878E+00	2.878E+00	2.878E+00
	IKE1	2.870E+00	2.870E+00	2.870E+00	2.870E+00	2.870E+00
	IKE2	2.875E+00	-	-	-	-
	JAE	2.879E+00	2.879E+00	2.878E+00	2.877E+00	2.877E+00
	PSI1	2.871E+00	2.871E+00	2.870E+00	2.870E+00	2.870E+00
	STU	2.861E+00	2.861E+00	2.861E+00	2.861E+00	2.861E+00
	Average	2.890E+00	2.872E+00	2.871E+00	2.871E+00	2.871E+00
Pu-240	BNFL	3.150E+00	3.150E+00	3.150E+00	3.150E+00	3.150E+00
	CEA	3.078E+00	3.081E+00	3.081E+00	3.080E+00	3.080E+00
	ECN	3.065E+00	3.064E+00	3.064E+00	3.063E+00	3.063E+00
	EDF	2.856E+00	3.125E+00	3.125E+00	3.125E+00	3.125E+00
	HIT	3.156E+00	3.155E+00	3.155E+00	3.155E+00	3.154E+00
	IKE1	3.076E+00	3.076E+00	3.075E+00	3.075E+00	3.075E+00
	IKE2	3.081E+00	-	-	-	-
	JAE	3.092E+00	3.091E+00	3.090E+00	3.089E+00	3.089E+00
	PSI1	3.081E+00	3.081E+00	3.079E+00	3.079E+00	3.078E+00
	STU	2.775E+00	2.775E+00	2.775E+00	2.775E+00	2.775E+00
	Average	3.041E+00	3.066E+00	3.066E+00	3.066E+00	3.066E+00
Pu-241	BNFL	2.937E+00	2.937E+00	2.936E+00	2.936E+00	2.936E+00
	CEA	2.937E+00	2.937E+00	2.936E+00	2.935E+00	2.936E+00
	ECN	2.946E+00	2.946E+00	2.943E+00	2.942E+00	2.943E+00
	EDF	3.125E+00	2.965E+00	2.964E+00	2.964E+00	2.964E+00
	HIT	2.937E+00	2.937E+00	2.936E+00	2.936E+00	2.936E+00
	IKE1	2.937E+00	2.937E+00	2.936E+00	2.936E+00	2.936E+00
	IKE2	2.937E+00	-	-	-	-
	JAE	2.936E+00	2.936E+00	2.935E+00	2.935E+00	2.935E+00
	PSI1	2.937E+00	2.937E+00	2.936E+00	2.936E+00	2.936E+00
	STU	2.917E+00	2.917E+00	2.917E+00	2.917E+00	2.917E+00
	Average	2.955E+00	2.939E+00	2.938E+00	2.937E+00	2.937E+00

Tab. 20: Neutrons per Fission, Benchmark B, (cont.)

Nuclide	Contributor	Burnup, MWd/kg				
		0.0	10.0	33.0	42.0	50.0
Pu-242	BNFL	3.106E+00	3.106E+00	3.106E+00	3.107E+00	3.107E+00
	CEA	3.119E+00	3.119E+00	3.120E+00	3.123E+00	3.121E+00
	ECN	3.105E+00	3.106E+00	3.105E+00	3.107E+00	3.107E+00
	EDF	3.581E+00	3.159E+00	3.159E+00	3.159E+00	3.159E+00
	HIT	3.145E+00	3.144E+00	3.143E+00	3.143E+00	3.144E+00
	IKE1	3.115E+00	3.115E+00	3.116E+00	3.116E+00	3.116E+00
	IKE2	3.119E+00	-	-	-	-
	JAE	3.123E+00	3.122E+00	3.122E+00	3.122E+00	3.122E+00
	PSI1	3.120E+00	3.120E+00	3.120E+00	3.120E+00	3.120E+00
	STU	2.808E+00	2.808E+00	2.808E+00	2.808E+00	2.808E+00
	Average	3.134E+00	3.089E+00	3.089E+00	3.089E+00	3.089E+00
Am-241	BNFL	3.565E+00	3.564E+00	3.556E+00	3.552E+00	3.548E+00
	CEA	-	3.570E+00	3.562E+00	3.557E+00	3.554E+00
	ECN	3.568E+00	3.566E+00	3.558E+00	3.554E+00	3.550E+00
	EDF	2.965E+00	3.581E+00	3.574E+00	3.570E+00	3.566E+00
	HIT	-	3.475E+00	3.468E+00	3.465E+00	3.461E+00
	IKE1	3.566E+00	3.565E+00	3.557E+00	3.553E+00	3.549E+00
	IKE2	-	-	-	-	-
	JAE	3.472E+00	3.470E+00	3.462E+00	3.458E+00	3.455E+00
	PSI1	3.573E+00	3.573E+00	3.565E+00	3.561E+00	3.557E+00
	STU	3.330E+00	3.330E+00	3.330E+00	3.330E+00	3.330E+00
	Average	3.434E+00	3.522E+00	3.515E+00	3.511E+00	3.508E+00
Am-242m	BNFL	2.701E+00	2.701E+00	2.701E+00	2.701E+00	2.701E+00
	CEA	-	3.211E+00	3.211E+00	3.211E+00	3.211E+00
	ECN	-	-	-	-	-
	EDF	3.252E+00	3.252E+00	3.252E+00	3.252E+00	3.252E+00
	HIT	-	-	-	-	-
	IKE1	3.211E+00	3.211E+00	3.211E+00	3.211E+00	3.211E+00
	IKE2	-	-	-	-	-
	JAE	3.276E+00	3.276E+00	3.276E+00	3.276E+00	3.275E+00
	PSI1	-	-	-	-	-
	STU	3.210E+00	3.210E+00	3.210E+00	3.210E+00	3.210E+00
	Average	3.130E+00	3.143E+00	3.143E+00	3.143E+00	3.143E+00

Tab. 20: Neutrons per Fission, Benchmark B, (cont.)

Nuclide	Contributor	Burnup, MWd/kg				
		0.0	10.0	33.0	42.0	50.0
Am-243	BNFL	3.476E+00	3.477E+00	3.479E+00	3.479E+00	3.479E+00
	CEA	-	3.490E+00	3.492E+00	3.492E+00	3.492E+00
	ECN	3.473E+00	3.475E+00	3.475E+00	3.475E+00	3.475E+00
	EDF	3.484E+00	3.485E+00	3.486E+00	3.487E+00	3.487E+00
	HIT	-	3.535E+00	3.539E+00	3.539E+00	3.540E+00
	IKE1	3.486E+00	3.487E+00	3.488E+00	3.489E+00	3.489E+00
	IKE2	-	-	-	-	-
	JAE	3.572E+00	3.572E+00	3.573E+00	3.573E+00	3.573E+00
	PSI1	3.497E+00	3.497E+00	3.494E+00	3.495E+00	3.495E+00
	STU	3.063E+00	3.063E+00	3.063E+00	3.063E+00	3.063E+00
	Average	3.436E+00	3.454E+00	3.454E+00	3.455E+00	3.455E+00
Cm-242	BNFL	-	-	-	-	-
	CEA	-	3.420E+00	3.413E+00	3.409E+00	3.406E+00
	ECN	3.408E+00	3.407E+00	3.400E+00	3.398E+00	3.394E+00
	EDF	3.417E+00	3.417E+00	3.416E+00	3.415E+00	3.415E+00
	HIT	-	3.790E+00	3.783E+00	3.780E+00	3.777E+00
	IKE1	3.416E+00	3.416E+00	3.409E+00	3.406E+00	3.403E+00
	IKE2	-	-	-	-	-
	JAE	3.477E+00	3.475E+00	3.471E+00	3.469E+00	3.467E+00
	PSI1	3.420E+00	3.420E+00	3.413E+00	3.410E+00	3.407E+00
	STU	3.150E+00	3.150E+00	3.150E+00	3.150E+00	3.150E+00
	Average	3.381E+00	3.437E+00	3.432E+00	3.430E+00	3.427E+00
Cm-243	BNFL	-	-	-	-	-
	CEA	-	3.397E+00	3.395E+00	3.395E+00	3.395E+00
	ECN	3.422E+00	3.423E+00	3.421E+00	3.421E+00	3.420E+00
	EDF	3.435E+00	3.435E+00	3.435E+00	3.435E+00	3.434E+00
	HIT	-	-	-	-	-
	IKE1	3.395E+00	3.395E+00	3.395E+00	3.395E+00	3.395E+00
	IKE2	-	-	-	-	-
	JAE	3.439E+00	3.439E+00	3.438E+00	3.438E+00	3.438E+00
	PSI1	3.396E+00	3.396E+00	3.395E+00	3.395E+00	3.395E+00
	STU	3.390E+00	3.390E+00	3.390E+00	3.390E+00	3.390E+00
	Average	3.413E+00	3.411E+00	3.410E+00	3.410E+00	3.410E+00

Tab. 20: Neutrons per Fission, Benchmark B, (cont.)

Nuclide	Contributor	Burnup, MWd/kg				
		0.0	10.0	33.0	42.0	50.0
Cm-244	BNFL	-	-	-	-	-
	CEA	-	3.525E+00	3.525E+00	3.525E+00	3.525E+00
	ECN	-	-	-	-	-
	EDF	-	-	-	-	-
	HIT	-	3.506E+00	3.513E+00	3.513E+00	3.513E+00
	IKE1	3.526E+00	3.527E+00	3.526E+00	3.526E+00	3.526E+00
	IKE2	-	-	-	-	-
	JAE	3.566E+00	3.562E+00	3.559E+00	3.558E+00	3.558E+00
	PSI1	-	-	-	-	-
	STU	3.240E+00	3.240E+00	3.240E+00	3.240E+00	3.240E+00
	Average	3.444E+00	3.472E+00	3.473E+00	3.473E+00	3.472E+00
Cm-245	BNFL	-	-	-	-	-
	CEA	-	-	-	-	-
	ECN	-	-	-	-	-
	EDF	-	-	-	-	-
	HIT	-	3.836E+00	3.835E+00	3.835E+00	3.834E+00
	IKE1	3.826E+00	3.826E+00	3.825E+00	3.825E+00	3.825E+00
	IKE2	-	-	-	-	-
	JAE	3.609E+00	3.609E+00	3.609E+00	3.608E+00	3.608E+00
	PSI1	-	-	-	-	-
	STU	3.820E+00	3.820E+00	3.820E+00	3.820E+00	3.820E+00
	Average	3.752E+00	3.773E+00	3.772E+00	3.772E+00	3.772E+00

Tab. 21: Nuclide Densities of Actinides, Benchmark B

Nuclide	Contributor	Burnup, MWd/kg				
		0.0	10.0	33.0	42.0	50.0
U-234	BNFL	2.463E -07	2.187E -07	1.625E -07	1.433E -07	1.274E -07
	CEA	2.463E -07	3.005E -07	4.069E -07	4.477E -07	4.866E -07
	ECN	2.462E -07	3.282E -07	4.554E -07	4.932E -07	5.242E -07
	EDF	2.463E -07	3.283E -07	4.421E -07	4.631E -07	4.726E -07
	HIT	2.463E -07	3.306E -07	4.656E -07	5.074E -07	5.426E -07
	IKE1	2.463E -07	3.293E -07	4.588E -07	4.968E -07	5.276E -07
	JAE	2.463E -07	3.299E -07	4.657E -07	5.090E -07	5.460E -07
	PSI1	2.463E -07	3.233E -07	4.416E -07	4.762E -07	5.044E -07
	STU	2.463E -07	3.289E -07	4.603E -07	4.990E -07	5.304E -07
	Average	2.463E -07	3.131E -07	4.177E -07	4.484E -07	4.735E -07
U-235	BNFL	5.152E -05	4.408E -05	2.902E -05	2.396E -05	1.989E -05
	CEA	5.152E -05	4.407E -05	2.891E -05	2.382E -05	1.972E -05
	ECN	5.151E -05	4.410E -05	2.907E -05	2.406E -05	2.007E -05
	EDF	5.152E -05	4.400E -05	2.870E -05	2.358E -05	1.950E -05
	HIT	5.152E -05	4.391E -05	2.854E -05	2.346E -05	1.942E -05
	IKE1	5.152E -05	4.412E -05	2.906E -05	2.402E -05	1.998E -05
	JAE	5.151E -05	4.393E -05	2.864E -05	2.358E -05	1.955E -05
	PSI1	5.151E -05	4.406E -05	2.896E -05	2.394E -05	1.994E -05
	STU	5.151E -05	4.413E -05	2.905E -05	2.398E -05	1.990E -05
	Average	5.151E -05	4.404E -05	2.888E -05	2.382E -05	1.977E -05
U-236	BNFL	-	1.720E -06	4.770E -06	5.621E -06	6.220E -06
	CEA	-	1.658E -06	4.589E -06	5.406E -06	5.981E -06
	ECN	-	1.680E -06	4.657E -06	5.486E -06	6.070E -06
	EDF	-	1.708E -06	4.723E -06	5.556E -06	6.136E -06
	HIT	-	1.748E -06	4.843E -06	5.700E -06	6.304E -06
	IKE1	-	1.719E -06	4.738E -06	5.563E -06	6.135E -06
	JAE	-	1.836E -06	5.033E -06	5.900E -06	6.498E -06
	PSI1	-	1.710E -06	4.728E -06	5.569E -06	6.159E -06
	STU	-	1.663E -06	4.593E -06	5.399E -06	5.961E -06
	Average	-	1.716E -06	4.742E -06	5.578E -06	6.163E -06

Tab. 21: Nuclide Densities of Actinides, Benchmark B, (cont.)

Nuclide	Contributor	Burnup, MWd/kg				
		0.0	10.0	33.0	42.0	50.0
U-238	BNFL	2.030E -02	2.016E -02	1.984E -02	1.970E -02	1.958E -02
	CEA	2.030E -02	2.016E -02	1.983E -02	1.969E -02	1.956E -02
	ECN	2.030E -02	2.016E -02	1.983E -02	1.968E -02	1.955E -02
	EDF	2.030E -02	2.016E -02	1.983E -02	1.969E -02	1.956E -02
	HIT	2.030E -02	2.016E -02	1.983E -02	1.969E -02	1.956E -02
	IKE1	2.030E -02	2.016E -02	1.984E -02	1.970E -02	1.957E -02
	JAE	2.029E -02	2.016E -02	1.983E -02	1.969E -02	1.956E -02
	PSI1	2.029E -02	2.016E -02	1.983E -02	1.969E -02	1.956E -02
	STU	2.029E -02	2.017E -02	1.984E -02	1.970E -02	1.958E -02
	Average	2.030E -02	2.016E -02	1.983E -02	1.969E -02	1.957E -02
Np-237	BNFL	-	8.445E -07	2.811E -06	3.483E -06	4.019E -06
	CEA	-	7.334E -07	2.359E -06	2.927E -06	3.383E -06
	ECN	-	7.991E -07	2.600E -06	3.212E -06	3.697E -06
	EDF	-	7.985E -07	2.557E -06	3.169E -06	3.660E -06
	HIT	-	8.891E -07	2.850E -06	3.495E -06	4.024E -06
	IKE1	-	6.884E -07	2.329E -06	2.918E -06	3.397E -06
	JAE	-	6.331E -07	2.098E -06	2.622E -06	3.045E -06
	PSI1	-	9.285E -07	3.011E -06	3.717E -06	4.275E -06
	STU	-	6.552E -07	2.150E -06	2.681E -06	3.118E -06
	Average	-	7.744E -07	2.529E -06	3.136E -06	3.624E -06
Pu-238	BNFL	2.180E -05	1.990E -05	1.610E -05	1.484E -05	1.381E -05
	CEA	2.180E -05	1.990E -05	1.820E -05	1.866E -05	1.943E -05
	ECN	2.179E -05	1.997E -05	1.908E -05	1.980E -05	2.067E -05
	EDF	2.178E -05	1.998E -05	1.908E -05	1.979E -05	2.065E -05
	HIT	2.180E -05	1.990E -05	1.875E -05	1.938E -05	2.020E -05
	IKE1	2.180E -05	1.999E -05	1.902E -05	1.968E -05	2.050E -05
	JAE	2.180E -05	2.000E -05	1.931E -05	2.020E -05	2.127E -05
	PSI1	2.180E -05	2.000E -05	1.890E -05	1.944E -05	2.014E -05
	STU	2.180E -05	1.999E -05	1.906E -05	1.975E -05	2.059E -05
	Average	2.180E -05	1.996E -05	1.861E -05	1.906E -05	1.970E -05

Tab. 21: Nuclide Densities of Actinides, Benchmark B, (cont.)

Nuclide	Contributor	Burnup, MWd/kg				
		0.0	10.0	33.0	42.0	50.0
Pu-239	BNFL	7.116E -04	5.908E -04	3.889E -04	3.335E -04	2.933E -04
	CEA	7.116E -04	5.886E -04	3.827E -04	3.264E -04	2.861E -04
	ECN	7.112E -04	5.921E -04	3.937E -04	3.399E -04	3.012E -04
	EDF	7.116E -04	5.894E -04	3.840E -04	3.289E -04	2.898E -04
	HIT	7.116E -04	5.900E -04	3.891E -04	3.345E -04	2.958E -04
	IKE1	7.116E -04	5.913E -04	3.874E -04	3.317E -04	2.917E -04
	JAE	7.116E -04	5.911E -04	3.914E -04	3.375E -04	2.988E -04
	PSI1	7.116E -04	5.958E -04	3.998E -04	3.469E -04	3.088E -04
	STU	7.116E -04	5.877E -04	3.799E -04	3.228E -04	2.832E -04
	Average	7.115E -04	5.907E -04	3.886E -04	3.336E -04	2.943E -04
Pu-240	BNFL	2.762E -04	2.888E -04	2.824E -04	2.697E -04	2.552E -04
	CEA	2.762E -04	2.845E -04	2.716E -04	2.574E -04	2.419E -04
	ECN	2.762E -04	2.843E -04	2.711E -04	2.573E -04	2.427E -04
	EDF	2.762E -04	2.833E -04	2.668E -04	2.513E -04	2.349E -04
	HIT	2.762E -04	2.862E -04	2.741E -04	2.604E -04	2.454E -04
	IKE1	2.762E -04	2.841E -04	2.710E -04	2.567E -04	2.413E -04
	JAE	2.762E -04	2.828E -04	2.664E -04	2.515E -04	2.359E -04
	PSI1	2.762E -04	2.844E -04	2.725E -04	2.594E -04	2.451E -04
	STU	2.762E -04	2.824E -04	2.639E -04	2.479E -04	2.308E -04
	Average	2.762E -04	2.845E -04	2.711E -04	2.568E -04	2.415E -04
Pu-241	BNFL	1.459E -04	1.563E -04	1.610E -04	1.560E -04	1.490E -04
	CEA	1.459E -04	1.593E -04	1.631E -04	1.562E -04	1.474E -04
	ECN	1.459E -04	1.587E -04	1.646E -04	1.590E -04	1.513E -04
	EDF	1.459E -04	1.611E -04	1.685E -04	1.624E -04	1.539E -04
	HIT	1.459E -04	1.587E -04	1.656E -04	1.601E -04	1.525E -04
	IKE1	1.459E -04	1.582E -04	1.629E -04	1.570E -04	1.491E -04
	JAE	1.459E -04	1.599E -04	1.663E -04	1.605E -04	1.525E -04
	PSI1	1.459E -04	1.576E -04	1.622E -04	1.568E -04	1.497E -04
	STU	1.459E -04	1.603E -04	1.665E -04	1.598E -04	1.510E -04
	Average	1.459E -04	1.589E -04	1.645E -04	1.586E -04	1.507E -04

Tab. 21: Nuclide Densities of Actinides, Benchmark B, (cont.)

Nuclide	Contributor	Burnup, MWd/kg				
		0.0	10.0	33.0	42.0	50.0
Pu-242	BNFL	4.764E -05	5.249E -05	6.698E -05	7.311E -05	7.839E -05
	CEA	4.764E -05	5.453E -05	7.611E -05	8.566E -05	9.413E -05
	ECN	4.763E -05	5.434E -05	7.522E -05	8.449E -05	9.274E -05
	EDF	4.764E -05	5.475E -05	7.814E -05	8.881E -05	9.833E -05
	HIT	4.764E -05	5.463E -05	7.618E -05	8.581E -05	9.431E -05
	IKE1	4.764E -05	5.469E -05	7.633E -05	8.584E -05	9.430E -05
	JAE	4.764E -05	5.513E -05	8.065E -05	9.238E -05	1.028E -04
	PSI1	4.764E -05	5.481E -05	7.629E -05	8.575E -05	9.411E -05
	STU	4.764E -05	5.522E -05	7.912E -05	8.996E -05	9.955E -05
	Average	4.764E -05	5.451E -05	7.611E -05	8.576E -05	9.430E -05
Am-241	BNFL	-	4.325E -06	9.732E -06	1.032E -05	1.027E -05
	CEA	-	3.309E -06	8.514E -06	9.506E -06	9.903E -06
	ECN	-	4.390E -06	9.900E -06	1.050E -05	1.045E -05
	EDF	-	4.468E -06	1.022E -05	1.084E -05	1.078E -05
	HIT	-	4.476E -06	1.041E -05	1.115E -05	1.118E -05
	IKE1	-	4.396E -06	9.906E -06	1.047E -05	1.038E -05
	JAE	-	4.480E -06	1.039E -05	1.109E -05	1.109E -05
	PSI1	-	4.237E -06	9.602E -06	1.020E -05	1.019E -05
	STU	-	4.427E -06	9.967E -06	1.049E -05	1.033E -05
	Average	-	4.279E -06	9.849E -06	1.051E -05	1.051E -05
Am-242m	BNFL	-	5.810E -08	2.052E -07	2.211E -07	2.188E -07
	CEA	-	4.255E -08	1.647E -07	1.853E -07	1.914E -07
	ECN	-	4.974E -08	1.707E -07	1.833E -07	1.812E -07
	EDF	-	5.918E -08	2.141E -07	2.308E -07	2.286E -07
	HIT	-	-	-	-	-
	IKE1	-	5.030E -08	1.760E -07	1.887E -07	1.862E -07
	JAE	-	7.887E -08	2.944E -07	3.220E -07	3.224E -07
	PSI1	-	-	-	-	-
	STU	-	5.493E -08	1.912E -07	2.033E -07	2.000E -07
	Average	-	5.624E -08	2.023E -07	2.192E -07	2.184E -07

Tab. 21: Nuclide Densities of Actinides, Benchmark B, (cont.)

Nuclide	Contributor	Burnup, MWd/kg				
		0.0	10.0	33.0	42.0	50.0
Am-243	BNFL	-	7.800E -06	2.275E -05	2.794E -05	3.231E -05
	CEA	-	6.292E -06	1.862E -05	2.299E -05	2.669E -05
	ECN	-	6.281E -06	1.856E -05	2.286E -05	2.647E -05
	EDF	-	6.447E -06	1.924E -05	2.385E -05	2.782E -05
	HIT	-	5.247E -06	1.616E -05	2.024E -05	2.384E -05
	IKE1	-	6.133E -06	1.839E -05	2.281E -05	2.655E -05
	JAE	-	6.178E -06	1.888E -05	2.359E -05	2.763E -05
	PSI1	-	6.126E -06	1.907E -05	2.378E -05	2.772E -05
	STU	-	5.510E -06	1.634E -05	2.013E -05	2.338E -05
	Average	-	6.224E -06	1.867E -05	2.313E -05	2.693E -05
Cm-242	BNFL	-	-	-	-	-
	CEA	-	3.540E -07	2.026E -06	2.540E -06	2.853E -06
	ECN	-	4.405E -07	2.273E -06	2.832E -06	3.187E -06
	EDF	-	4.395E -07	2.299E -06	2.874E -06	3.242E -06
	HIT	-	4.062E -07	2.182E -06	2.756E -06	3.147E -06
	IKE1	-	4.370E -07	2.265E -06	2.829E -06	3.194E -06
	JAE	-	4.650E -07	2.487E -06	3.131E -06	3.553E -06
	PSI1	-	4.054E -07	2.086E -06	2.600E -06	2.910E -06
	STU	-	4.453E -07	2.341E -06	2.926E -06	3.304E -06
	Average	-	4.241E -07	2.245E -06	2.811E -06	3.174E -06
Cm-243	BNFL	-	-	-	-	-
	CEA	-	3.381E -09	5.896E -08	9.079E -08	1.172E -07
	ECN	-	3.767E -09	6.269E -08	9.647E -08	1.255E -07
	EDF	-	5.638E -09	9.415E -08	1.456E -07	1.902E -07
	HIT	-	-	-	-	-
	IKE1	-	3.717E -09	6.015E -08	9.224E -08	1.195E -07
	JAE	-	3.711E -09	6.487E -08	1.012E -07	1.329E -07
	PSI1	-	3.523E -09	5.613E -08	8.564E -08	1.103E -07
	STU	-	3.788E -09	6.414E -08	9.914E -08	1.297E -07
	Average	-	3.932E -09	6.587E -08	1.016E -07	1.322E -07

Tab. 21: Nuclide Densities of Actinides, Benchmark B, (cont.)

Nuclide	Contributor	Burnup, MWd/kg				
		0.0	10.0	33.0	42.0	50.0
Cm-244	BNFL	-	-	-	-	-
	CEA	-	9.346E -07	8.572E -06	1.312E -05	1.775E -05
	ECN	-	-	-	-	-
	EDF	-	-	-	-	-
	HIT	-	7.447E -07	7.195E -06	1.119E -05	1.541E -05
	IKE1	-	8.891E -07	8.366E -06	1.295E -05	1.767E -05
	JAE	-	8.926E -07	8.316E -06	1.281E -05	1.743E -05
	PSI1	-	-	-	-	-
	STU	-	8.217E -07	7.679E -06	1.180E -05	1.608E -05
	Average	-	8.565E -07	8.026E -06	1.238E -05	1.687E -05
Cm-245	BNFL	-	-	-	-	-
	CEA	-	-	-	-	-
	ECN	-	-	-	-	-
	EDF	-	-	-	-	-
	HIT	-	2.098E -08	5.599E -07	9.943E -07	1.477E -06
	IKE1	-	2.358E -08	6.146E -07	1.112E -06	1.659E -06
	JAE	-	2.959E -08	7.986E -07	1.445E -06	2.151E -06
	PSI1	-	-	-	-	-
	STU	-	2.641E -08	6.860E -07	1.228E -06	1.834E -06
	Average	-	2.514E -08	6.648E -07	1.195E -06	1.780E -06

Tab. 22: Nuclide Densities of Fiss. Prod., Benchmark B

Nuclide	Contributor	Burnup, MWd/kg				
		0.0	10.0	33.0	42.0	50.0
Mo-95	BNFL	-	1.078E -05	3.436E -05	4.317E -05	5.081E -05
	CEA	-	4.106E -06	2.709E -05	3.592E -05	4.345E -05
	ECN	-	5.407E -06	2.906E -05	3.806E -05	4.583E -05
	EDF	-	-	-	-	-
	HIT	-	5.254E -06	2.798E -05	3.656E -05	4.394E -05
	IKE1	-	5.089E -06	2.724E -05	3.556E -05	4.269E -05
	JAE	-	1.035E -05	3.288E -05	4.124E -05	4.846E -05
	PSI1	-	4.917E -06	2.676E -05	3.492E -05	4.189E -05
	STU	-	-	-	-	-
	Average	-	6.556E -06	2.934E -05	3.792E -05	4.530E -05
Tc-99	BNFL	-	1.285E -05	4.030E -05	5.028E -05	5.879E -05
	CEA	-	1.287E -05	4.055E -05	5.040E -05	5.866E -05
	ECN	-	1.178E -05	3.604E -05	4.436E -05	5.128E -05
	EDF	-	-	-	-	-
	HIT	-	1.306E -05	4.105E -05	5.103E -05	5.942E -05
	IKE1	-	1.291E -05	4.042E -05	5.011E -05	5.821E -05
	JAE	-	1.295E -05	4.073E -05	5.067E -05	5.905E -05
	PSI1	-	1.269E -05	3.954E -05	4.901E -05	5.695E -05
	STU	-	-	-	-	-
	Average	-	1.273E -05	3.980E -05	4.941E -05	5.748E -05
Ru-101	BNFL	-	-	-	-	-
	CEA	-	1.326E -05	4.264E -05	5.365E -05	6.319E -05
	ECN	-	-	-	-	-
	EDF	-	-	-	-	-
	HIT	-	1.291E -05	4.146E -05	5.218E -05	6.148E -05
	IKE1	-	1.273E -05	4.097E -05	5.155E -05	6.073E -05
	JAE	-	1.280E -05	4.112E -05	5.174E -05	6.094E -05
	PSI1	-	-	-	-	-
	STU	-	-	-	-	-
	Average	-	1.292E -05	4.155E -05	5.228E -05	6.158E -05

Tab. 22: Nuclide Densities of Fiss. Prod., Benchmark B,(cont.)

Nuclide	Contributor	Burnup, MWd/kg				
		0.0	10.0	33.0	42.0	50.0
Rh-103	BNFL	-	-	-	-	-
	CEA	-	9.934E -06	3.518E -05	4.248E -05	4.781E -05
	ECN	-	-	-	-	-
	EDF	-	-	-	-	-
	HIT	-	1.073E -05	3.529E -05	4.236E -05	4.750E -05
	IKE1	-	1.045E -05	3.439E -05	4.128E -05	4.633E -05
	JAE	-	1.069E -05	3.517E -05	4.223E -05	4.741E -05
	PSI1	-	-	-	-	-
	STU	-	1.394E -05	3.863E -05	4.570E -05	5.075E -05
	Average	-	1.115E -05	3.573E -05	4.281E -05	4.796E -05
Pd-105	BNFL	-	-	-	-	-
	CEA	-	1.143E -05	3.725E -05	4.696E -05	5.538E -05
	ECN	-	-	-	-	-
	EDF	-	-	-	-	-
	HIT	-	1.110E -05	3.580E -05	4.501E -05	5.293E -05
	IKE1	-	1.106E -05	3.585E -05	4.517E -05	5.323E -05
	JAE	-	1.108E -05	3.588E -05	4.517E -05	5.320E -05
	PSI1	-	-	-	-	-
	STU	-	-	-	-	-
	Average	-	1.116E -05	3.620E -05	4.558E -05	5.369E -05
Pd-107	BNFL	-	-	-	-	-
	CEA	-	7.339E -06	2.406E -05	3.048E -05	3.609E -05
	ECN	-	-	-	-	-
	EDF	-	-	-	-	-
	HIT	-	7.591E -06	2.481E -05	3.141E -05	3.717E -05
	IKE1	-	7.350E -06	2.387E -05	3.010E -05	3.550E -05
	JAE	-	7.557E -06	2.482E -05	3.147E -05	3.730E -05
	PSI1	-	-	-	-	-
	STU	-	-	-	-	-
	Average	-	7.459E -06	2.439E -05	3.087E -05	3.651E -05

Tab. 22: Nuclide Densities of Fiss. Prod., Benchmark B,(cont.)

Nuclide	Contributor	Burnup, MWd/kg				
		0.0	10.0	33.0	42.0	50.0
Pd-108	BNFL	-	-	-	-	-
	CEA	-	4.909E -06	1.660E -05	2.129E -05	2.548E -05
	ECN	-	-	-	-	-
	EDF	-	-	-	-	-
	HIT	-	5.153E -06	1.735E -05	2.223E -05	2.659E -05
	IKE1	-	5.136E -06	1.746E -05	2.245E -05	2.693E -05
	JAE	-	5.109E -06	1.721E -05	2.208E -05	2.644E -05
	PSI1	-	-	-	-	-
	STU	-	-	-	-	-
	Average	-	5.077E -06	1.716E -05	2.201E -05	2.636E -05
Ag-109	BNFL	-	3.055E -06	8.757E -06	1.056E -05	1.198E -05
	CEA	-	3.271E -06	8.960E -06	1.065E -05	1.196E -05
	ECN	-	2.008E -06	5.079E -06	5.858E -06	6.414E -06
	EDF	-	-	-	-	-
	HIT	-	3.270E -06	9.273E -06	1.117E -05	1.269E -05
	IKE1	-	3.539E -06	9.606E -06	1.136E -05	1.268E -05
	JAE	-	3.559E -06	9.754E -06	1.159E -05	1.301E -05
	PSI1	-	3.212E -06	8.990E -06	1.082E -05	1.228E -05
	STU	-	3.131E -06	8.063E -06	9.295E -06	1.009E -05
	Average	-	3.131E -06	8.560E -06	1.016E -05	1.139E -05
Xe-131	BNFL	-	7.228E -06	1.977E -05	2.328E -05	2.583E -05
	CEA	-	7.030E -06	1.963E -05	2.299E -05	2.532E -05
	ECN	-	6.227E -06	1.707E -05	1.997E -05	2.203E -05
	EDF	-	-	-	-	-
	HIT	-	7.114E -06	2.039E -05	2.425E -05	2.710E -05
	IKE1	-	6.841E -06	1.865E -05	2.168E -05	2.375E -05
	JAE	-	6.943E -06	2.021E -05	2.433E -05	2.754E -05
	PSI1	-	7.344E -06	2.001E -05	2.358E -05	2.618E -05
	STU	-	7.356E -06	1.944E -05	2.256E -05	2.466E -05
	Average	-	7.010E -06	1.940E -05	2.283E -05	2.530E -05

Tab. 22: Nuclide Densities of Fiss. Prod., Benchmark B,(cont.)

Nuclide	Contributor	Burnup, MWd/kg				
		0.0	10.0	33.0	42.0	50.0
Xe-135	BNFL	-	-	-	-	-
	CEA	-	2.078E -08	1.569E -08	1.402E -08	1.269E -08
	ECN	-	-	-	-	-
	EDF	-	-	-	-	-
	HIT	-	1.816E -08	1.550E -08	1.442E -08	1.359E -08
	IKE1	-	1.862E -08	1.576E -08	1.464E -08	1.367E -08
	JAE	-	1.862E -08	1.592E -08	1.490E -08	1.397E -08
	PSI1	-	-	-	-	-
	STU	-	1.803E -08	1.537E -08	1.415E -08	1.338E -08
	Average	-	1.884E -08	1.565E -08	1.443E -08	1.346E -08
Cs-133	BNFL	-	1.433E -05	4.359E -05	5.367E -05	6.201E -05
	CEA	-	1.410E -05	4.378E -05	5.384E -05	6.207E -05
	ECN	-	1.288E -05	3.907E -05	4.774E -05	5.460E -05
	EDF	-	-	-	-	-
	HIT	-	1.415E -05	4.417E -05	5.453E -05	6.307E -05
	IKE1	-	1.402E -05	4.311E -05	5.289E -05	6.082E -05
	JAE	-	1.420E -05	4.440E -05	5.497E -05	6.379E -05
	PSI1	-	1.386E -05	4.286E -05	5.278E -05	6.094E -05
	STU	-	1.442E -05	4.317E -05	5.274E -05	6.042E -05
	Average	-	1.400E -05	4.302E -05	5.289E -05	6.097E -05
Cs-135	BNFL	-	-	-	-	-
	CEA	-	7.878E -06	2.497E -05	3.123E -05	3.660E -05
	ECN	-	-	-	-	-
	EDF	-	-	-	-	-
	HIT	-	8.882E -06	2.686E -05	3.291E -05	3.783E -05
	IKE1	-	9.031E -06	2.731E -05	3.353E -05	3.865E -05
	JAE	-	9.112E -06	2.775E -05	3.417E -05	3.951E -05
	PSI1	-	-	-	-	-
	STU	-	8.781E -06	2.637E -05	3.219E -05	3.686E -05
	Average	-	8.737E -06	2.665E -05	3.280E -05	3.789E -05

Tab. 22: Nuclide Densities of Fiss. Prod., Benchmark B,(cont.)

Nuclide	Contributor	Burnup, MWd/kg				
		0.0	10.0	33.0	42.0	50.0
Nd-143	BNFL	-	9.745E -06	2.973E -05	3.649E -05	4.194E -05
	CEA	-	8.700E -06	2.831E -05	3.469E -05	3.965E -05
	ECN	-	9.696E -06	3.152E -05	3.883E -05	4.461E -05
	EDF	-	-	-	-	-
	HIT	-	1.158E -05	3.894E -05	4.853E -05	5.627E -05
	IKE1	-	8.749E -06	2.811E -05	3.446E -05	3.944E -05
	JAE	-	8.890E -06	2.835E -05	3.477E -05	3.982E -05
	PSI1	-	8.772E -06	2.800E -05	3.432E -05	3.930E -05
	STU	-	9.612E -06	2.887E -05	3.510E -05	3.993E -05
	Average	-	9.468E -06	3.023E -05	3.715E -05	4.262E -05
Nd-145	BNFL	-	6.727E -06	2.089E -05	2.593E -05	3.018E -05
	CEA	-	6.741E -06	2.103E -05	2.612E -05	3.039E -05
	ECN	-	7.684E -06	2.449E -05	3.061E -05	3.580E -05
	EDF	-	-	-	-	-
	HIT	-	6.771E -06	2.139E -05	2.670E -05	3.122E -05
	IKE1	-	6.677E -06	2.086E -05	2.592E -05	3.016E -05
	JAE	-	6.738E -06	2.118E -05	2.639E -05	3.081E -05
	PSI1	-	6.722E -06	2.099E -05	2.609E -05	3.040E -05
	STU	-	6.777E -06	2.106E -05	2.612E -05	3.033E -05
	Average	-	6.855E -06	2.149E -05	2.674E -05	3.116E -05
Pm-147	BNFL	-	1.634E -06	2.996E -06	3.117E -06	3.144E -06
	CEA	-	3.402E -06	7.131E -06	7.537E -06	7.630E -06
	ECN	-	3.682E -06	7.714E -06	8.299E -06	8.581E -06
	EDF	-	-	-	-	-
	HIT	-	3.519E -06	7.445E -06	8.036E -06	8.324E -06
	IKE1	-	3.393E -06	6.875E -06	7.287E -06	7.442E -06
	JAE	-	3.451E -06	7.306E -06	7.875E -06	8.145E -06
	PSI1	-	3.415E -06	7.107E -06	7.649E -06	7.907E -06
	STU	-	3.628E -06	6.945E -06	7.317E -06	7.416E -06
	Average	-	3.265E -06	6.690E -06	7.140E -06	7.324E -06

Tab. 22: Nuclide Densities of Fiss. Prod., Benchmark B,(cont.)

Nuclide	Contributor	Burnup, MWd/kg				
		0.0	10.0	33.0	42.0	50.0
Pm-148m	BNFL	-	5.098E -08	9.620E -08	9.957E -08	9.935E -08
	CEA	-	5.720E -08	1.074E -07	1.060E -07	1.011E -07
	ECN	-	5.267E -08	1.042E -07	1.088E -07	1.079E -07
	EDF	-	-	-	-	-
	HIT	-	5.356E -08	1.112E -07	1.156E -07	1.196E -07
	IKE1	-	5.360E -08	1.083E -07	1.120E -07	1.109E -07
	JAE	-	5.948E -08	1.153E -07	1.205E -07	1.198E -07
	PSI1	-	5.399E -08	1.096E -07	1.129E -07	1.118E -07
	STU	-	5.970E -08	1.149E -07	1.164E -07	1.160E -07
	Average	-	5.515E -08	1.084E -07	1.115E -07	1.108E -07
Sm-149	BNFL	-	3.514E -07	3.152E -07	2.899E -07	2.643E -07
	CEA	-	2.491E -07	2.224E -07	2.030E -07	1.855E -07
	ECN	-	2.302E -07	2.069E -07	1.937E -07	1.804E -07
	EDF	-	-	-	-	-
	HIT	-	2.463E -07	2.207E -07	2.042E -07	1.931E -07
	IKE1	-	2.558E -07	2.316E -07	2.155E -07	2.004E -07
	JAE	-	2.560E -07	2.336E -07	2.205E -07	2.064E -07
	PSI1	-	2.550E -07	2.347E -07	2.210E -07	2.068E -07
	STU	-	2.550E -07	2.275E -07	2.065E -07	1.934E -07
	Average	-	2.623E -07	2.366E -07	2.193E -07	2.038E -07
Sm-150	BNFL	-	2.719E -06	1.075E -05	1.389E -05	1.659E -05
	CEA	-	2.709E -06	1.033E -05	1.326E -05	1.575E -05
	ECN	-	2.479E -06	9.298E -06	1.194E -05	1.422E -05
	EDF	-	-	-	-	-
	HIT	-	2.732E -06	1.033E -05	1.334E -05	1.597E -05
	IKE1	-	2.683E -06	1.018E -05	1.312E -05	1.566E -05
	JAE	-	2.696E -06	1.019E -05	1.316E -05	1.577E -05
	PSI1	-	2.745E -06	1.079E -05	1.407E -05	1.697E -05
	STU	-	2.718E -06	1.004E -05	1.280E -05	1.510E -05
	Average	-	2.685E -06	1.024E -05	1.320E -05	1.575E -05

Tab. 22: Nuclide Densities of Fiss. Prod., Benchmark B,(cont.)

Nuclide	Contributor	Burnup, MWd/kg				
		0.0	10.0	33.0	42.0	50.0
Sm-151	BNFL	-	1.225E -06	2.001E -06	2.035E -06	2.026E -06
	CEA	-	8.952E -07	1.153E -06	1.156E -06	1.151E -06
	ECN	-	7.371E -07	9.208E -07	9.274E -07	9.322E -07
	EDF	-	-	-	-	-
	HIT	-	9.015E -07	1.101E -06	1.089E -06	1.079E -06
	IKE1	-	9.119E -07	1.176E -06	1.184E -06	1.184E -06
	JAE	-	9.091E -07	1.161E -06	1.163E -06	1.158E -06
	PSI1	-	8.592E -07	1.004E -06	9.861E -07	9.733E -07
	STU	-	9.028E -07	1.182E -06	1.189E -06	1.193E -06
	Average	-	9.177E -07	1.212E -06	1.216E -06	1.212E -06
Sm-152	BNFL	-	1.604E -06	5.288E -06	6.339E -06	7.095E -06
	CEA	-	1.788E -06	5.188E -06	5.987E -06	6.548E -06
	ECN	-	1.429E -06	3.945E -06	4.474E -06	4.845E -06
	EDF	-	-	-	-	-
	HIT	-	1.884E -06	5.920E -06	6.995E -06	7.757E -06
	IKE1	-	1.857E -06	5.746E -06	6.734E -06	7.444E -06
	JAE	-	1.902E -06	6.336E -06	7.657E -06	8.675E -06
	PSI1	-	1.775E -06	4.853E -06	5.475E -06	5.871E -06
	STU	-	1.830E -06	5.372E -06	6.236E -06	6.812E -06
	Average	-	1.759E -06	5.331E -06	6.237E -06	6.881E -06
Eu-153	BNFL	-	9.943E -07	4.487E -06	6.023E -06	7.329E -06
	CEA	-	1.127E -06	5.374E -06	7.106E -06	8.512E -06
	ECN	-	8.642E -07	4.045E -06	5.295E -06	6.287E -06
	EDF	-	-	-	-	-
	HIT	-	1.068E -06	4.789E -06	6.301E -06	7.546E -06
	IKE1	-	1.065E -06	5.008E -06	6.694E -06	8.108E -06
	JAE	-	1.053E -06	4.624E -06	6.114E -06	7.371E -06
	PSI1	-	1.184E -06	5.482E -06	7.124E -06	8.387E -06
	STU	-	1.166E -06	5.428E -06	7.175E -06	8.614E -06
	Average	-	1.065E -06	4.905E -06	6.479E -06	7.769E -06

Tab. 22: Nuclide Densities of Fiss. Prod., Benchmark B,(cont.)

Nuclide	Contributor	Burnup, MWd/kg				
		0.0	10.0	33.0	42.0	50.0
Eu-154	BNFL	-	1.277E -07	1.609E -06	2.628E -06	3.660E -06
	CEA	-	1.308E -07	1.557E -06	2.421E -06	3.223E -06
	ECN	-	9.967E -08	1.170E -06	1.810E -06	2.396E -06
	EDF	-	-	-	-	-
	HIT	-	1.233E -07	1.192E -06	1.732E -06	2.193E -06
	IKE1	-	1.211E -07	1.396E -06	2.193E -06	2.955E -06
	JAE	-	1.171E -07	1.144E -06	1.689E -06	2.159E -06
	PSI1	-	1.347E -07	1.604E -06	2.475E -06	3.266E -06
	STU	-	1.361E -07	1.566E -06	2.429E -06	3.248E -06
	Average	-	1.238E -07	1.405E -06	2.172E -06	2.888E -06
Eu-155	BNFL	-	1.534E -07	2.729E -07	3.653E -07	4.655E -07
	CEA	-	2.204E -07	5.264E -07	7.048E -07	8.684E -07
	ECN	-	1.557E -07	3.723E -07	5.078E -07	6.335E -07
	EDF	-	-	-	-	-
	HIT	-	1.145E -07	2.936E -07	4.094E -07	5.092E -07
	IKE1	-	2.094E -07	4.853E -07	6.499E -07	8.071E -07
	JAE	-	1.130E -07	2.716E -07	3.719E -07	4.686E -07
	PSI1	-	2.730E -07	5.960E -07	7.774E -07	9.432E -07
	STU	-	2.827E -07	6.017E -07	7.769E -07	9.473E -07
	Average	-	1.903E -07	4.275E -07	5.704E -07	7.054E -07

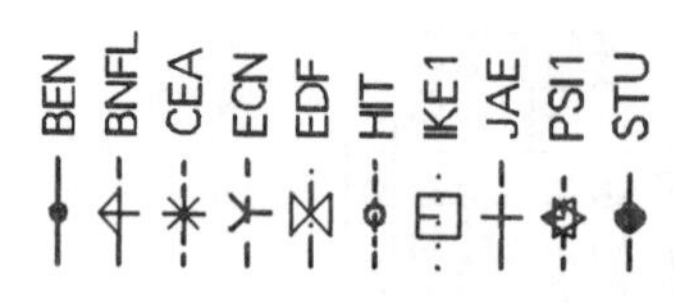

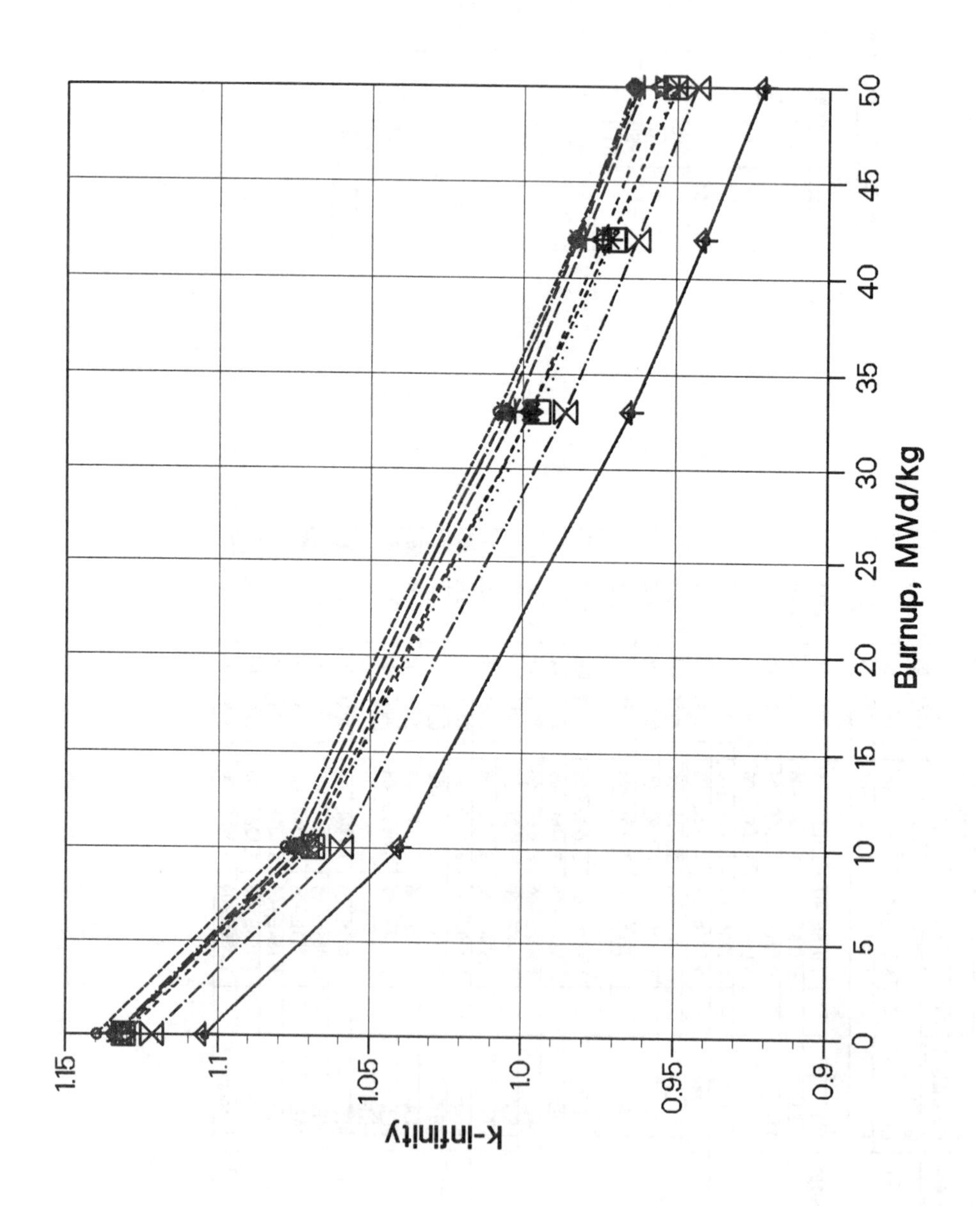

Fig. 1-A: k-infinity Benchmark A

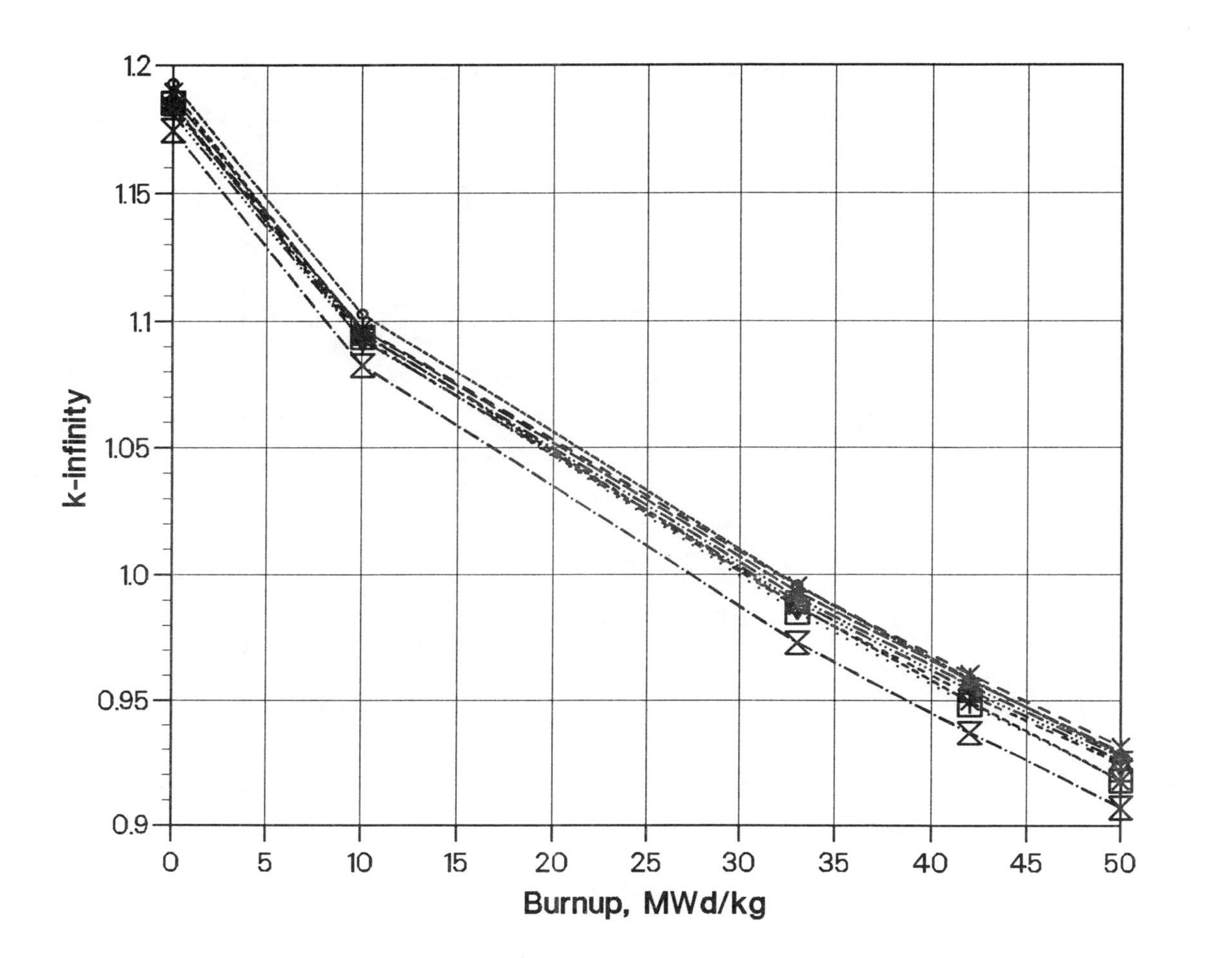

Fig. 1-B: k-infinity Benchmark B

Appendix A

Benchmark specification for plutonium recycling in PWRs

Benchmark A: Poor-quality plutonium
J. Vergnes (EDF)

Benchmark B: Better plutonium vector
H. W. Wiese (KfK) and G. Schlosser (Siemens-KWU)

Co-ordinator
H. Küsters, KfK

Benchmark A – poor-isotopic-quality plutonium

The goal of this comparison is to explain the reasons for unexplained differences between results on MOX-PWR cell calculations using degraded plutonium (fifth-stage recycle).

The most important difference is related to the infinite medium multiplication constant k-infinity. We suggest a geometry as simple as possible. We shall describe the proposed options:

- Number of atoms and cell geometry

  Differences could appear for these calculations. So we propose that a number of atoms will be stated for the benchmark.

  For this preliminary calculation, we have taken the geometry of Figure A-1 and the following isotopic balance of plutonium. The plutonium isotopic composition is near the composition at the fifth stage recycle with an average burnup of 50 MWd/kg.

Pu-238	4%
Pu-239	36%
Pu-240	28%
Pu-241	12%
Pu-242	20%

The uranium isotopic composition is the following

U-235	0.711%
U-238	99.289%

The total plutonium concentration proposed is 12.5% (6% of fissile plutonium).

The cladding is only made out of natural zirconium.

In evolution, samarium and xenon concentrations will be self-estimated by each code with a nominal power of 38.3 W/g of initial heavy metal.

- Options of the cell calculation

To ease the comparisons, it is suggested to calculate the cell without any neutron leakage ($B^2 = 0$).

Temperatures will be as follows:

- Fuel 660°C
- Cladding 306.3°C
- Water 306.3°C

Boron concentration is worth 500 ppm. Boron composition is as follows:

- B-10 18.3%
- B-11 81.7%

FUEL	
	ATOMS / cm^3
U-234	0
U-235	$1.4456 \cdot 10^{20}$
U-236	0
U-238	$1.9939 \cdot 10^{22}$
Np-237	0
Pu-238	$1.1467 \cdot 10^{20}$
Pu-239	$1.0285 \cdot 10^{21}$
Pu-240	$7.9657 \cdot 10^{20}$
Pu-241	$3.3997 \cdot 10^{20}$
Pu-242	$5.6388 \cdot 10^{20}$
Am-241	0
Am-242	0
Am-243	0
Cm-242	0
Cm-243	0
CLADDING	
natural Zr	$4.5854 \cdot 10^{22}$
MODERATOR	
H_2O	$2.3858 \cdot 10^{22}$
B-10	$3.6346 \cdot 10^{18}$
B-11	$1.6226 \cdot 10^{19}$

Table A-1 ***Number of atoms per cm³ at irradiation step zero***

- Options of the evolution calculation

We propose an evolution calculation from 0 to 50 MWd/kg including the following time steps (0, 0.15, 0.5, 1, 2, 4, 6, 10, 15, 20, 22, 26, 30, 33, 38, 42, 47 and 50 MWd/kg)

We take into consideration the following fission products:

Zr-95, Mo-95, Pd-106, Ce-144, Pm-147, Pm-148, Pm-148m, Sm-149, Sm-150, Sm-151, Sm-152, Eu-153, Eu-154, Eu-155, Gd-155, Gd-156, Gd-157, Tc-99, Ag-109, Cd-113, In-115, I-129, Xe-131, Cs-131, Cs-137, Nd-143, Nd-145, Nd-148
and four pseudo fission products in which all the other fission products are grouped.

The energy releases from fission are:

NUCLIDE	ENERGY RELEASE (MeV)
U-235	193.7
U-238	197.0
Pu-239	202.0
Pu-241	204.4
Am-242m	207.0

plus 8 MeV for the n-gamma captures of the other non-fissioning (ν-1) neutrons.

- Results

Results should be provided both on paper and computer-processable medium. A short report should be provided describing:

- The computer program(s) used and their precise version,
- The data libraries used and evaluated data file from which they were derived,
- The list of isotopes for which resonance self-shielding was applied and the method used,
- How the buildup of Xenon was treated,
- How the (n,2n)-reaction was taken into account for the k-infinity calculation.

The following data should be provided in tabular form for the following burnups:
0, 10, 33, 42 and 50 MWd/kg.

1. Number densities for all nuclides considered:

	burnup 1	burnup 2	....	burnup-n
isotope 1				
isotope 2				
.				
.				
.				
.				
.				
isotope -N				

2. k as a function of burnup,
3. One energy group cross-section (absorption, fission, nu-bar) as a function of isotope and burnup (see 1.),
4. Reaction rates (absorption, fission) as a function of isotope and burnup (see 1.),
5. Applied absolute fluxes used in the evolution calculation (and their normalisation factor),

6. Neutron energy spectrum per unit lethargy as a function of burnup (and its normalisation factor and group structure).

Benchmark B – better plutonium vector

As a second fuel M2, in agreement both with Dr. G. Schlosser, KWU and Dr. J. Vergnes, EDF, a MOX fuel with first-generation-plutonium as used in [1] with the following specifications is suggested:

- 4.0 wt% U-235 in uranium tailings (0.25 wt% U-235),
- Composition of plutonium (wt%):

Pu-238	1.8
Pu-239	59.0
Pu-240	23.0
Pu-241	12.2
Pu-242	4.0

- Composition of uranium (wt%):

U-234	0.00119
U-235	0.25
U-238	99.74881

With the heavy material number density normalised to 2.115×10^{22} atoms /cm^3, the following nuclide number densities are determined:

NUCLIDE	ATOMS / cm^3
U-234	$2.4626 \cdot 10^{17}$
U-235	$5.1515 \cdot 10^{19}$
U-238	$2.0295 \cdot 10^{22}$
Pu-238	$2.1800 \cdot 10^{19}$
Pu-239	$7.1155 \cdot 10^{20}$
Pu-240	$2.7623 \cdot 10^{20}$
Pu-241	$1.4591 \cdot 10^{20}$
Pu-242	$4.7643 \cdot 10^{19}$
heavy metal-atoms	$2.155 \cdot 10^{22}$
O	$4.310 \cdot 10^{22}$

All other specifications shall be the same as in the first benchmark – case A.

Reference

[1] H. W. Wiese, "Investigation of the Nuclear Inventories of High-Exposure PWR Mixed Oxide Fuels with Multiple Recycling of Self-Generating Plutonium", Nuclear Technology, Vol. 102, April 1993, p. 68.

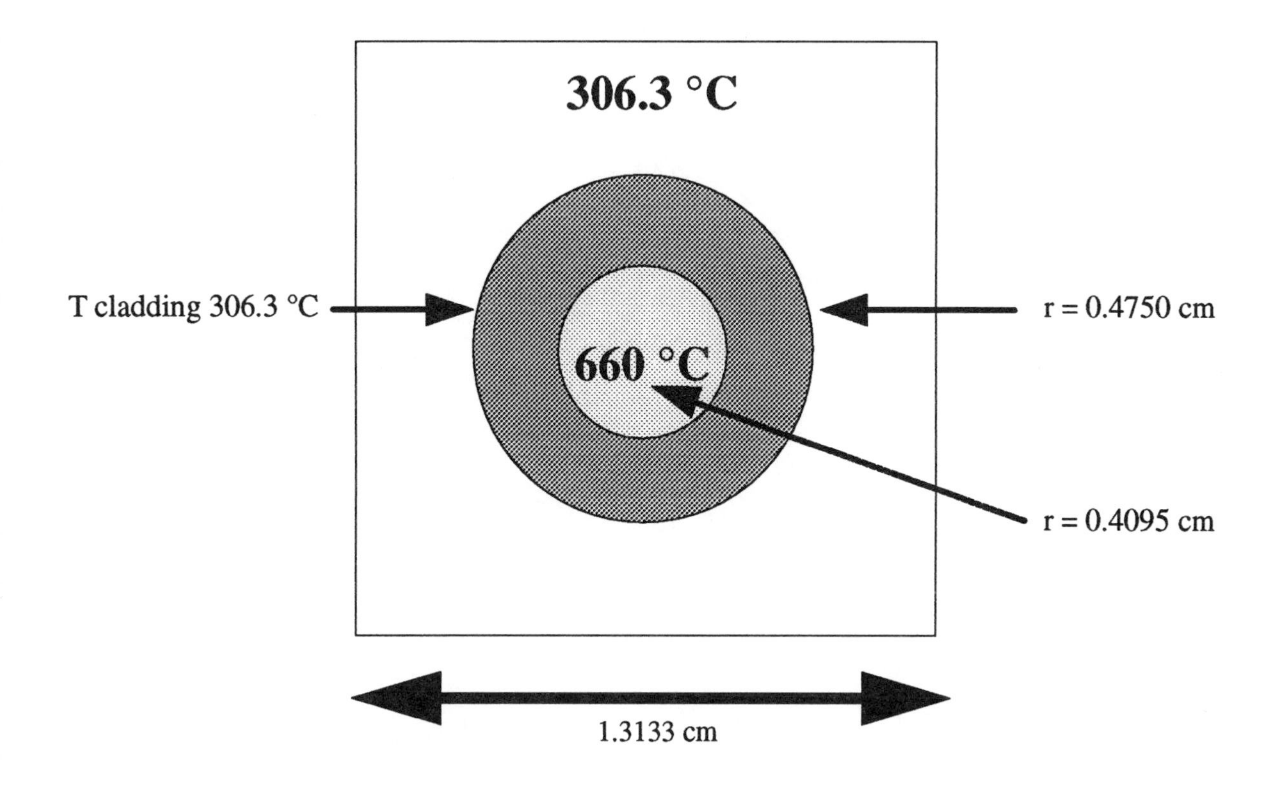

Figure A-1 ***Cell geometry at 20°C***

Appendix B.1

VIM Monte Carlo calculations

R. N. Blomquist (ANL)

General

Generally, the continuous-energy cross-section data used was based on ENDF/B-V data. Zircaloy, however, was based on ENDF/B-IV. Specular reflection was applied at all boundary surfaces. The uncertainties shown in these results are one standard deviation of the mean. The flux and fission rate data submitted is normalised to per fission source neutron, and volume-integrated over each cell. Isotopic multigroup cross-sections and reaction rates are available on request.

Plutonium recycling in PWR benchmark

Both calculations consisted of 0.5 million histories each in generations of 10 000, with tally data written for each generation. Two initial generations were discarded to allow the flat fission source guess to decay. Computation time averaged 5 SPARC-2 CPU hours.

```
MOX          Tue Dec 21 14:20:50 1993          1

The following are the VIM Monte Carlo results for the two MOX pincell
benchmark problems specified in NSC/DOC(93)19.

Roger Blomquist                 708-252-8423
Reactor Analysis Division       708-252-4500(FAX)
Argonne National Laboratory     RNBlomquist@anl.gov (internet)
9700 S. Cass Ave.
Argonne, IL 60439

--------------------------------------------------------------------------------
--------------------------------------------------------------------------------

MOX pincell NSC/DOC(93)19

--------------------------------------------------------------------------------
BOC Cell Problem A (Dirty Plutonium) Results - 300 degrees(K)

1.  Atom Densities (at/cc); same as problem specifications

Fuel:
U-235  1.445600E+20
U-238  1.993900E+22
Pu-238 1.146700E+20
Pu-239 1.028500E+21
Pu-240 7.965700E+20
Pu-241 3.399700E+20
Pu-242 5.638800E+20
O-16   4.585100E+22

Clad:
Zirc2  4.324800E+22

Moderator:
H      4.771600E+22
O      2.385800E+22
B-10   3.634600E+18
B-11   1.622600E+19


2.  K-eff:  1.1591 +/- 0.0011

3.  Microscopic One-Group Cross Sections:

Nuclide      Absorption (uncertainty)    Fission (uncertainty)     Nu-bar
Fuel:
U-235        1.7157E+01 (1.92E-01%)      1.2685E+01 (1.85E-01%)    2.4462
U-238        8.6690E-01 (1.79E-01%)      1.2119E-01 (2.25E-01%)    2.8157
Pu-238       1.0360E+01 (2.93E-01%)      1.8656E+00 (1.84E-01%)    3.0449
Pu-239       3.1222E+01 (1.61E-01%)      2.0164E+01 (1.54E-01%)    2.8946
Pu-240       2.0524E+01 (2.45E-01%)      6.5058E-01 (1.18E-01%)    3.1177
Pu-241       3.6387E+01 (1.57E-01%)      2.7204E+01 (1.61E-01%)    2.9587
Pu-242       6.5597E+00 (5.17E-01%)      5.0944E-01 (1.38E-01%)    3.1574
O            3.6205E-03 (6.27E-01%)
Clad:
Zirc2        3.8853E-02 (6.40E-01%)

Moderator:
H            9.8391E-03 (2.79E-01%)
O            3.3921E-03 (5.03E-01%)
B-10         1.1381E+02 (2.79E-01%)
B-11         1.8229E-04 (5.78E-01%)


4.  Isotopic Reaction Rates:
(per fission source neutron, integrated over E and vol

Nuclide      Absorption (uncertainty)      Fission (uncertainty)
Fuel:
U-235        2.83878E-02 (2.40E-01%)       2.09845E-02 (2.09E-01%)
U-238        1.96458E-01 (2.89E-01%)       2.73787E-02 (5.49E-01%)
Pu-238       1.35368E-02 (5.47E-01%)       2.44071E-03 (6.29E-01%)
Pu-239       3.64803E-01 (1.31E-01%)       2.35761E-01 (1.39E-01%)
Pu-240       1.85927E-01 (2.56E-01%)       5.90460E-03 (5.00E-01%)
Pu-241       1.41212E-01 (1.69E-01%)       1.05543E-01 (1.59E-01%)
Pu-242       4.19104E-02 (5.18E-01%)       3.27896E-03 (5.43E-01%)
O            1.86476E-03 (1.38E+00%)

Clad:
Zirc2        6.64000E-03 (1.86E+00%)

Moderator:
H            1.01717E-02 (1.03E+00%)
O            1.80119E-03 (3.64E+00%)
B-10         8.96309E-03 (1.02E+00%)
B-11         6.02591E-08 (2.26E+00%)

Totals:      1.00168E+00 (5.59E-03%)       4.01291E-01 (9.71E-02%)

5.  One-group flux in fuel: 1.1374E+01 +/- 7.80E-02% cm/fission source neutron
(integrated over E and volume)

6.  Flux spectrum by region (Emax=20MeV)
(per fission source neutron, integrated over E and volume)

Fuel:
 GROUP  ELOWER(EV)     FLUX       FLUX/LETH   Flux Uncertainty
   1    8.21000E+05 3.6195E+00  1.1336E+00  2.17E-01%
   2    5.53000E+03 4.5347E+00  9.0687E-01  1.65E-01%)
   3    6.25000E-01 2.9606E+00  3.2578E-01  1.14E-01%)
   4    1.00000E-05 2.5003E-01  2.2642E-02  2.35E-01%)
 TOTALS             1.1365E+01              9.27E-02%)

Clad:
 GROUP  ELOWER(EV)     FLUX       FLUX/LETH   Flux Uncertainty
   1    8.21000E+05 1.2041E+00  3.7712E-01  2.48E-01%)
   2    5.53000E+03 1.5434E+00  3.0866E-01  2.49E-01%)
   3    6.25000E-01 1.0794E+00  1.1877E-01  2.85E-01%)
   4    1.00000E-05 1.2685E-01  1.1487E-02  8.00E-01%)
 TOTALS             3.9538E+00              1.77E-01%)

Moderator:
 GROUP  ELOWER(EV)     FLUX       FLUX/LETH   Flux Uncertainty
   1    8.21000E+05 6.4964E+00  2.0346E+00  1.81E-01%)
   2    5.53000E+03 8.4667E+00  1.6932E+00  7.81E-02%)
   3    6.25000E-01 6.2073E+00  6.8302E-01  8.99E-02%)
   4    1.00000E-05 8.5354E-01  7.7293E-02  3.11E-01%)
 TOTALS             2.2024E+01              7.60E-02%)

--------------------------------------------------------------------------------
BOC Cell Problem B (Better Plutonium) Results - 300 degrees(K)

1.  Atom Densities (at/cc); same as problem specifications

Fuel:
U-234  2.462600E+17
U-235  5.151500E+19
U-238  2.029500E+22
Pu-238 2.180000E+19
Pu-239 7.115500E+20
Pu-240 2.762300E+20
```

```
Pu-241 1.459100E+20
Pu-242 4.764300E+19
O-16   4.310000E+22

Clad:
Zirc2  4.324800E+22

Moderator:
H      4.771600E+22
O      2.385800E+22
B-10   3.634600E+18
B-11   1.622600E+19


2.  K-eff:  1.2117 +/- 0.0010

3.  One-group Microscopic Cross Sections:

Nuclide      Absorption (uncertainty)     Fission (uncertainty)      Nu-bar
Fuel:
U-234        2.0754E+01 (1.70E+00%)       5.7432E-01 (1.28E-01%)     2.6264
U-235        2.5748E+01 (1.72E-01%)       1.9892E+01 (1.69E-01%)     2.4425
U-238        9.0724E-01 (1.84E-01%)       1.1595E-01 (2.34E-01%)     2.8154
Pu-238       2.0149E+00 (2.08E-01%)       1.6745E+01 (2.63E-01%)     3.0276
Pu-239       5.2908E+01 (1.41E-01%)       3.4507E+01 (1.39E-01%)     2.8909
Pu-240       4.1560E+01 (3.17E-01%)       6.2834E-01 (1.31E-01%)     3.1142
Pu-241       5.8969E+01 (1.33E-01%)       4.3619E+01 (1.32E-01%)     2.9565
Pu-242       2.1007E+01 (8.84E-01%)       4.8854E-01 (1.41E-01%)     3.1563
O            3.4214E-03 (7.48E-01%)

Clad:
Zirc2        4.1421E-02 (5.56E-01%)

Moderator:
H            1.5029E-02 (1.59E-01%)
O            3.2037E-03 (4.98E-01%)
B-10         1.7378E+02 (1.59E-01%)
B-11         2.5638E-04 (3.46E-01%)


4.  Isotopic Reaction Rates:
(per fission source neutron, integrated over E and volume)

Nuclide      Absorption (uncertainty)     Fission (uncertainty)
Fuel:
U-234        6.14499E-05 (4.07E+00%)      1.72942E-06 (8.33E-01%)
U-235        1.61353E-02 (2.56E-01%)      1.24676E-02 (2.20E-01%)
U-238        2.24294E-01 (2.18E-01%)      2.86300E-02 (8.50E-01%)
Pu-238       4.43677E-03 (5.82E-01%)      5.36310E-04 (8.35E-01%)
Pu-239       4.57851E-01 (9.33E-02%)      2.98592E-01 (8.97E-02%)
Pu-240       1.39490E-01 (2.82E-01%)      2.11970E-03 (8.29E-01%)
Pu-241       1.04684E-01 (1.65E-01%)      7.74376E-02 (1.50E-01%)
Pu-242       1.20382E-02 (1.01E+00%)      2.83645E-04 (8.52E-01%)
O            1.77790E-03 (1.79E+00%)

Clad:
Zirc2        7.28400E-03 (1.66E+00%)

Moderator:
H            1.69191E-02 (7.84E-01%)
O            1.79336E-03 (3.10E+00%)
B-10         1.49014E-02 (7.82E-01%)
B-11         1.54158E-07 (3.98E+01%)
```

```
Totals:      1.00167E+00 (5.80E-03%)     4.20068E-01 (8.92E-02%)

5.  One-group flux in fuel:  1.2146+1 +/- 8.6E-2% cm/fission source neutron
(integrated over E and volume)

6.  Flux spectrum by region (Emax=20MeV)
(per fission source neutron, integrated over E and volume)

Fuel:
 Group  Elower(EV)     Flux        Flux/Leth    Flux Uncertainty
   1    8.21000E+05 3.7224E+00  1.1658E+00   (2.36E-01%)
   2    5.53000E+03 4.6034E+00  9.2062E-01   (1.40E-01%)
   3    6.25000E-01 3.3225E+00  3.6559E-01   (1.61E-01%)
   4    1.00000E-05 5.1820E-01  4.6926E-02   (2.38E-01%)
 TOTALS             1.2166E+01               (1.13E-01%)

Clad:
 Group  Elower(EV)     Flux        Flux/Leth    Flux Uncertainty
   1    8.21000E+05 1.2413E+00  3.8875E-01   (2.41E-01%)
   2    5.53000E+03 1.5649E+00  3.1296E-01   (2.14E-01%)
   3    6.25000E-01 1.1977E+00  1.3179E-01   (2.59E-01%)
   4    1.00000E-05 2.2700E-01  2.0556E-02   (7.19E-01%)
 TOTALS             4.2309E+00               (1.48E-01%)

Moderator:
 Group  Elower(EV)     Flux        Flux/Leth    Flux Uncertainty
   1    8.21000E+05 6.7136E+00  2.1026E+00   (1.58E-01%)
   2    5.53000E+03 8.6066E+00  1.7212E+00   (7.54E-02%)
   3    6.25000E-01 6.8074E+00  7.4906E-01   (8.63E-02%)
   4    1.00000E-05 1.4494E+00  1.3125E-01   (2.60E-01%)
 TOTALS             2.3577E+01               (7.02E-02%)

--------------------------------------------------------------------------------
--------------------------------------------------------------------------------
```

Appendix B.2

Recycling of plutonium in a PWR

A. Puill (CEA) and A. Kolmayer (Framatome)

The evolution of the infinite multiplication factor and the isotopic concentrations are carried out with the **APOLLO-2** code [1] to [4].

APOLLO-2 is a modular code which solves the multigroup transport equation, either by the method of collision probabilities (integral equation), or by S_n methods with finite differences or nodal techniques (integro-differential equation).

It can process 1-D or 2-D geometries (a multicell approximation is available for 2-D geometries). APOLLO-2 is a portable code written in FORTRAN 77.

The neutronic data appear in one or several external libraries which have a format designed for the code. There are two types of data: isotopic data and self-shielding data. The latter are used by the self-shielding module, which calculates self-shielded cross-sections in order to make complete the isotopic data in the resonance domain.

A series of tests enabled us to select the best calculation options taking into account the **accuracy** of the result as well as the computation **time**. The **PIJ** option (collision probabilities) is used in **rectangular** multicell geometry (flat flux approximation in each of the regions) and the **UP1** approximation is used to represent the incoming and outgoing angular fluxes (Uniform in space and P1 – 3 terms – in angle). The quadrature formulae used in the calculation of the transmission of probabilities, are of Gauss-Legendre type.

The self-shielding module calculates the self-shielded multigroup cross-sections for all the resonant isotopes located in the multicell geometry. The calculation takes into account the resonant interaction effects in space and energy between the various isotopic mixtures .

A test has been carried out with an exact 2-D-PIJ method for the calculation of fluxes and self-shielded cross-sections. With regard to the adopted options, the k-infinity deviation does not exceed 0.05% for a computation time 65 times larger.

In order to take into account the high flux gradient in the fuel, it is divided into 6 rings with decreasing thicknesses from the centre to the periphery.

The CEA 93 library uses an energy mesh with 172 groups ranging from 0 to 20 MeV. Most of the isotopes are coming from the JEF-2.2 evaluated data library. The quadrature formulae used for the self-shielding calculation are supplied for 7 heavy isotopes: U-235, U-238, Pu-238, Pu-239, Pu-240, Pu-241, Pu-242 and for the natural zirconium. The self-shielded sections are recalculated every 10 MWd/kg.

The xenon is saturated at the first depletion step. The contribution of the (n, 2n)-reactions is taken into account in the calculation of k-infinity (in APOLLO-2, for a cell without leakage, we have k-infinity = k-effective). Moreover these multigroup cross-sections are weighted by a flux calculated with 172 groups.

Acknowledgements

The first author wishes to thank the colleagues of the "APOLLO Team": A. Constans, M. Coste, M. C. Laigle, G. Mathonnière and H. Tellier for their kind help and support during this work.

References

[1] H. Tellier, M. Coste, C. Raepsaet and C. Van der Gucht, "Heavy Nucleus Resonant Absorption in Heterogeneous Lattices. II: Physical Qualification", Note CEA-N-2701.

[2] M. Coste, H. Tellier, P. Ribon, C. Raepsaet and C. Van der Gucht, "New Improvements in the Self-Shielding Formalism of the APOLLO-2 Code", Mathematical Methods and Supercomputing in Nuclear Applications, April 19th - 23rd, 1993, Karlsruhe, Germany.

[3] D. Belhaffaf, M. Coste, R. Lenain, G. Mathonnière, R. Sanchez, Z. Stankovski and I. Zmijarevic "Use of the APOLLO-2 Transport Code for PWR Assembly Studies", Topical meeting on Advances in Reactor Physics, March 8th - 11th, 1992, Charleston, U.S.A.

[4] R. Sanchez, J. Mondot, Z. Stankovski, A. Cossic and I. Zmijarevic, "APOLLO-2: A User-Oriented, Portable Modular Code for Multigroup Transport Assembly Calculations", International Conference on Development of Reactor Physics and Calculation Methods. April 27th - 30th, 1987, Paris, France.

OECD/NEA Benchmark A on Plutonium recycling in PWRs

APOLLO-2 Calculation options effects on the infinite multiplication factor

March 1994

BURNUP	LIBRARY CEA 86/93	1R/6R	99g/172g	WITHOUT / WITH *	UP0/UP1 **	(n,2n) WITHOUT / WITH **	*TOTAL*
0	-163	-410	-261	0	-379	-170	-1383
10	-306	-310	-50	+140	–	–	-1075
20	-413	-266	+34	+186	–	–	-1008
30	-423	-200	+108	+227	–	–	-837
40	-293	-102	+199	+272	–	–	-473
50	-119	-24	+253	+289	–	–	-150

* Recalculation of self-shielding.

** Test made only at 0 MWd/kg

Appendix B.3

Results of OECD/WPPR benchmark on plutonium recycling in PWRs

V. A. Wichers and J. M. Li (ECN)

General

Computer programs

The calculations were done with the following program sequence:

- **BONAMI** Calculates resonance self-shielding in the unresolved region based on the Bondarenko method;

- **NITAWL** Calculates resonance self-shielding in the resolved region based on the Nordheim method;

- **WIMS-D** (version 4). Produce spectrally and spatially weighted (collapsed) cross-sections for the point-depletion computations, and k-infinity and the flux for the single cell;

- **COUPLE** Produces an ORIGEN-S nuclear data library, using the cross-section library and spectra produced by WIMS-D.
- Modifies the ORIGEN-S spectral parameters (THERM, RES and FAST);

- **ORIGEN-S** Calculates fuel composition as a function of burnup from a point-depletion computation.

- **SAS6** Absorption in nuclides not specified in the benchmark was approximately accounted for in the flux calculations. These nuclides were treated as 1/v-absorbers. Appropriate concentrations of these "pseudo fission-products" were computed for three energy intervals: 1.0E-5 eV – 0.5 eV; 0.5 eV – 1 MeV; and 1 MeV – 20 MeV.

All codes were from the SCALE version 4.1 package, except for WIMS-D.

Data libraries

With respect to cross-section computation, nuclides specified in the benchmark and nuclides not specified were treated differently.

For nuclides specified in the benchmark, the cross-section data library was an AMPX master library with the 172 groups XMAS structure. This library was based on the JEF-2.2 evaluated data file for all nuclides, with the exception of O-16, Gd-155 and the 1/v-absorbers, for which the JEF-1.1 evaluated data file was used. The master library was generated with the processing code AJAX of the SCALE system, version 4.1.

For nuclides not specified in the benchmark, the ORIGEN-S cross-section data library was used.

Self-shielding

Resonance self-shielding was in principle applied to all nuclides explicitly specified in the benchmark, and to the following nuclides:

Kr-83, Zr-93, M-97, Mo-98, Ru-101, Ru-103, Rh-103, Pd-107, Pd-108, I-127, Xe-135, Cs-134, Cs-135, La-139, Pr-141, Sm-147.

Pd-105 was not included because of problems with the nuclear data.

Resonance self-shielding was not applied to other nuclides.

Build-up of xenon

The xenon concentrations were self estimated. There was no special treatment of the build-up of xenon.

Treatment of (n.2n)-reactions in the k-infinity calculations

In WIMS-D, (n,xn)-reactions were taken into account as negative absorption reactions, by using appropriately modified absorption cross sections.

Cell geometry

The cell geometry was as given in Figure A-1 of the benchmark specification (see Appendix A).

The length of the cell was 10^7 cm, in order to have effectively no neutron leakage i.e.,

$$DB^2 = 0.$$

Boron concentrations

The boron concentration was constant during the computations.

Fission energies

In the benchmark, energy releases from fission were specified for five nuclides:

U-235, U-238, Pu-239, Pu-241, and Am-242m.

For the remaining fissioning nuclides, we used the fission energies included in the ORIGEN-S code.

A fixed energy release of 8 MeV was specified for the n-gamma captures for the other, (ν-1), non-fissioning neutrons. We interpreted this as: the energy releases from capture reactions of the (ν-1) non-fissioning neutrons sum to 8 MeV. Thus, all fission energies were incremented by 8 MeV, and all capture energies in ORIGEN-S were set to zero.

The adopted energies released per fission are listed in Table B.3-1.

NUCLIDE	ATOMIC MASS (g/mol)	ENERGY (MeV)
Th-230		198.0
Th-232		197.21
Th-233		198.0
Pa-231		198.0
Pa-233		197.1
U-232		208.0
U-233		199.29
U-234	234.114	198.30
U-235	235.04401	201.70
U-236		200.80
U-238	238.05099	205.00
Np-237		203.10
Pu-238	238.21344	205.8
Pu-239	239.13	210.0
Pu-240	240.054	207.79
Pu-241	241.05685	212.4
Pu-242	242.05847	208.62
Pu-243		208.0
Am-241		210.3
Am-242m		215.0
Am-243		210.1
Cm-244		208.0
Cm-245		208.0

Table B.3-1 ***Adopted total energies generated per fission and relevant atomic masses of actinides.***

Flux normalisation

The ORIGEN-S depletion computations were normalised to the power generated in the fuel pin. This power was defined by the benchmark specification through the power density, being 38.3 W/g of initial heavy metal. The power corresponding to this 38.3 W/g of initial heavy metal was obtained from the specified initial atom densities, the dimensions of the pin and atomic masses (see Table B.3-1).

For part A (poor-quality plutonium case) the power was 1830.5 MW, the mass density was 9.0722 g/cm^3, and the volume was 5.2681.10^6 cm^3 (length of 10^7 cm, fuel radius of 0.4095 cm).

For part B (better plutonium case) the power was 1719.4 MW, the mass density was 8.5215 g/cm^3, and the volume was again 5.2681.10^6 cm^3.

The flux F' used by ORIGEN was converted to the total flux F by:

$$F = F'/C$$
$$C = 1/(1 + RES.\ln(2.10^6) + FAST/1.45).$$

Pseudo fission products

ORIGEN updates densities for all nuclides.

For nuclides specified in the benchmark, microscopic cross-sections corrected for resonance self-shielding were computed by BONAMI and NITAWL, using the updated densities from ORIGEN and microscopic cross-sections from the master library. These updated microscopic cross-sections were used in the subsequent flux computations with WIMS-D.

Nuclides not specified in the benchmark were approximately and partially taken into account in the flux computations as 1/v absorbers. Appropriate densities were computed for three energy regions.

Number densities

An initial density of 10^6 atoms/cm^3 was used as effective density = 0 atoms/cm^3.

One energy group cross-sections

All cross-sections are in units of barns. Cross-sections S' computed by COUPLE, were converted to one-energy-group cross-sections, S, by:

$$S = C \times S'$$
$$C = 1/(1 + RES.\ln(2.10^6) + FAST/1.45).$$

Appendix B.4

Plutonium Recycling in PWRs

P. Marimbeau (CEA) and P. Barbrault and J. Vergnes (EDF)

The calculations are performed with the code APOLLO-1 [1] [2].

This code solves the multigroup transport equation with the method of collision probabilities (PIJ). It is portable and written in FORTRAN 77.

The CEA 86.1 [3] library is used with 99 energy-groups ranging from 0 to 10 MeV.

The self-shielded cross-sections are recalculated at each step of burnup for the heavy isotopes:

U-235, U-236, U-238, Pu-239, Pu-240, Pu-241, Pu-242, and natural zirconium.

The xenon is not saturated at the first step of irradiation (0 MWd/kg). The equilibrium is obtained at 0.15 MWd/kg.

The contribution of the (n,2n)-reactions is not taken into account in the calculation of k-infinity. It is taken into account in the k-effective.

The fission rates are normalised in such a way that the total absorption rate is 1. plus the (n,2n)-contribution. Moreover, the total production rate equals k-effective, which is k-infinity plus the (n,2n)-contribution.

The neutron energy spectrum (99 groups) is an average over the whole calculation cell.

Acknowledgements

The second author (P. Barbrault) thanks his colleagues E. Vial, J. Vergnes, C. Bangil and S. Marguet for their kind help during this work.

References

[1] A. Kavenoky, "APOLLO: A General Code for Transport, Slowing-Down and Thermalization Calculations in Heterogeneous Media".
Proc. National Topical Meeting on Mathematical Models and Computational Techniques for Analysis of Nuclear Systems, Ann Arbor, Michigan, April 9-11, 1973.

[2] A. Kavenoky and R. Sanchez, "The APOLLO 1 Assembly Spectrum Code".
ENS Topical Meeting on Advances in Reactors Physics, Mathematics and Computation, Paris, 1987.

[3] A. Santamarina and H. Tellier, "The French CEA-86 Multigroup Cross-Section Library and its Integral Qualification".
Proc. of International Conference on Nuclear Data for Science and Technology, May 30 - June 1988, MITO, Japan.

Appendix B.5

Benchmark calculation of a fuel assembly analysis code VMONT for PWR-MOX lattice

K. Ishii and H. Maruyama (Hitachi Ltd)

Introduction

This report describes the PWR-MOX benchmark results by using a fuel assembly analysis code VMONT [1] [2], in which a multigroup Monte Carlo neutron transport calculation is combined with a burnup calculation. The algorithm of this Monte Carlo calculation was developed for effective use of the vector processing function of supercomputers such as Hitachi S-820.

Calculational model

Multigroup cross-section library

The total number of energy groups used in the spectrum calculation of the VMONT code is 190, the structure of which is shown in Table B.5-1. Infinite dilute cross-sections and self-shielding factors are stored in a multigroup cross-section library, and the self-shielding factors are tabulated as a function of background cross-sections and temperatures.

The multigroup cross-section library is prepared mainly on the basis of the JENDL-2 and the ENDF/B-IV nuclear data files. Table B.5-2 shows the data base of the principal nuclides. The fast and epithermal group cross-sections are processed with the MINX code [3] and the thermal group cross-sections are provided with the FLANGE-IV code [4].

Neutron spectrum calculation

The VMONT code calculates the neutron spectra using the vectorized Monte Carlo neutron transport method. The basic features of the Monte Carlo method used in this code are:

1. a multi-particle tracking algorithm suited to the vector processing ability of Hitachi supercomputers;
2. a pseudoscattering scheme [5] used in the flight analysis, and;
3. a "zone sampling" method [2] used in the zone identification of collision sites.

Owing to these features, the VMONT code can realise speeds more than 20 times faster than those of a scalar Monte Carlo code.

The VMONT code considers the (n,2n)-reaction as the negative absorption in the k-infinity calculation.

Burnup calculation

The VMONT code treats 138 nuclides, including 32 actinides and 84 fission products as shown in Table B.5-3. The actinide and fission product chains used in the burnup calculation are shown Figures B.5-1 and B.5-2. Total fission yield of the all fission products treated explicitly in the code is 0.9 to 1.1 depending on the fissile nuclides. Other fission products are collapsed into one lumped fission product.

The effective few-group cross sections used in the burnup calculation are generated in each burnup step through condensation of the 190-group cross sections where the self-shielding effects are taken into account for the important fission products as well as the actinides.

Results

The statistical uncertainties of the VMONT code are as follows:

- k-infinity = 0.05%
- Flux, microscopic cross section, and reaction rate = 0.2%
- Relative fission rate distribution = 1%

The fluxes and reaction rates are normalised to total production.

References

[1] Y. Morimoto, et al., "Neutronic Analysis Code for Fuel Assembly Using a Vectorized Monte Carlo Method", Nucl. Sci. Eng., 103, 351 (1989).

[2] H. Maruyama, et al., "Development and Performance Evaluation of a Vectorized Monte Carlo Method with Pseudoscattering", Proceedings of the First International Conference on Supercomputing in Nuclear Applications, 156 (1990).

[3] C. R Weisbin, et al, "MINX: A Multigroup Interpretation of Nuclear X-sections from ENDF/B", LA-6486-MS, Los Alamos National Laboratory (1976).

[4] H C. Honeck and D. R Finch, "FLANGE: A Code to Process Thermal Neutron Data from an ENDF/B Tape", Atomic Energy Commission Research and Development Report, DP-1278. E I. Du Pont de Nemours and Company (1973).

[5] E. R Woodcock, et al., Proc. Conf. on Methods to Reactor Problems, ANL-7050 (1965).

Energy Range		Groups
Fast and Epithermal	10.0MeV to 0.683eV	132 (Equi-Lethargy)
Thermal	0.683eV to 0.0eV	58

Table B.5-1 ***Energy group structure of the VMONT code***

Material	JENDL-2	ENDF/B-IV
Fuel	^{228}Th, ^{233}U, ^{234}U, ^{236}U, ^{237}Np, ^{239}Np, ^{236}Pu, ^{238}Pu, ^{241}Am, ^{242}Am, ^{243}Am, ^{242}Cm, ^{244}Cm, ^{245}Cm	^{235}U, ^{238}U, ^{239}Pu, ^{240}Pu, ^{241}Pu, ^{242}Pu
Fission Product	^{83}Kr, ^{93}Zr, ^{95}Mo, ^{97}Mo, ^{98}Mo, ^{99}Tc, ^{101}Ru, ^{103}Rh, ^{105}Pd, ^{107}Pd, ^{108}Pd, ^{109}Ag, ^{113}Cd, ^{129}I, ^{131}Xe, ^{133}Xe, ^{135}Xe, ^{133}Cs, ^{135}Cs, ^{139}La, ^{141}Pr, ^{143}Nd, ^{144}Nd, ^{145}Nd, ^{148}Nd, ^{147}Pm, ^{147}Sm, ^{148}Sm, ^{149}Sm, ^{150}Sm, ^{151}Sm, ^{152}Sm, ^{154}Sm, ^{153}Eu, ^{154}Eu, ^{155}Eu, ^{154}Gd, ^{155}Gd, ^{156}Gd, ^{157}Gd, ^{158}Gd	^{100}Ru, ^{105}Rh, ^{143}Pr, ^{148m}Pm, ^{148}Pm, ^{149}Pm, ^{156}Eu
Structure Control Moderator	Al, ^{12}C, ^{10}B, ^{1}H	Zr, Fe, ^{16}O, ^{11}B

Table B.5-2 ***Content of principal nuclides in the VMONT library***

Material	Number of Nuclides
Fuel	32
Fission Product	83 + 1 Lumped Fission Product
Structure, Control, Moderator	22

Table B.5-3 ***Number of nuclides treated in the VMONT code***

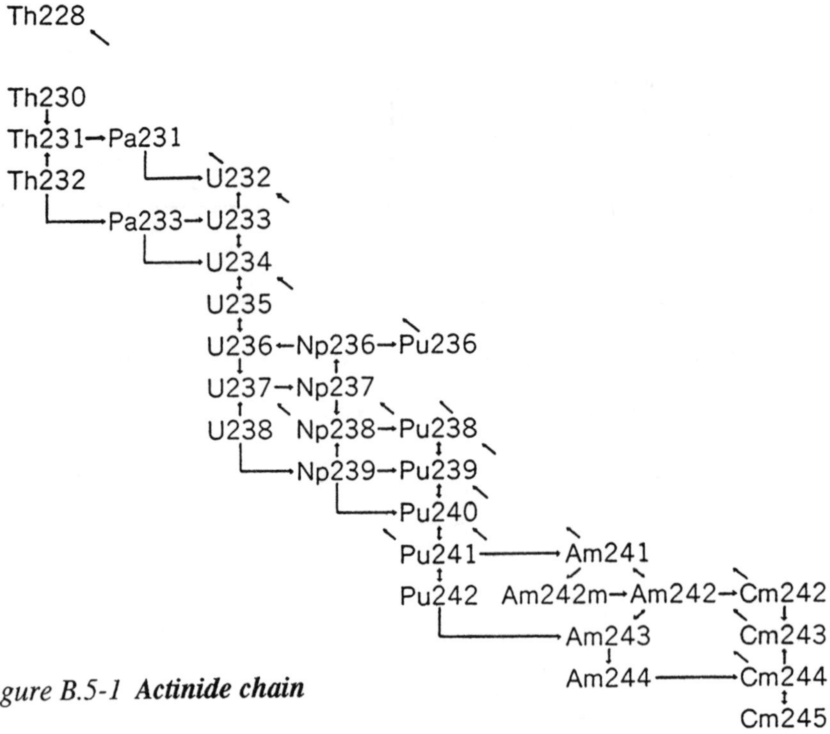

Th228

Th230
Th231 → Pa231
Th232 U232
Pa233 → U233
U234
U235
U236 ← Np236 → Pu236
U237 → Np237
U238 Np238 → Pu238
Np239 → Pu239
Pu240
Pu241 → Am241
Pu242 Am242m → Am242 → Cm242
Am243 Cm243
Am244 → Cm244
Cm245

Figure B.5-1 **Actinide chain**

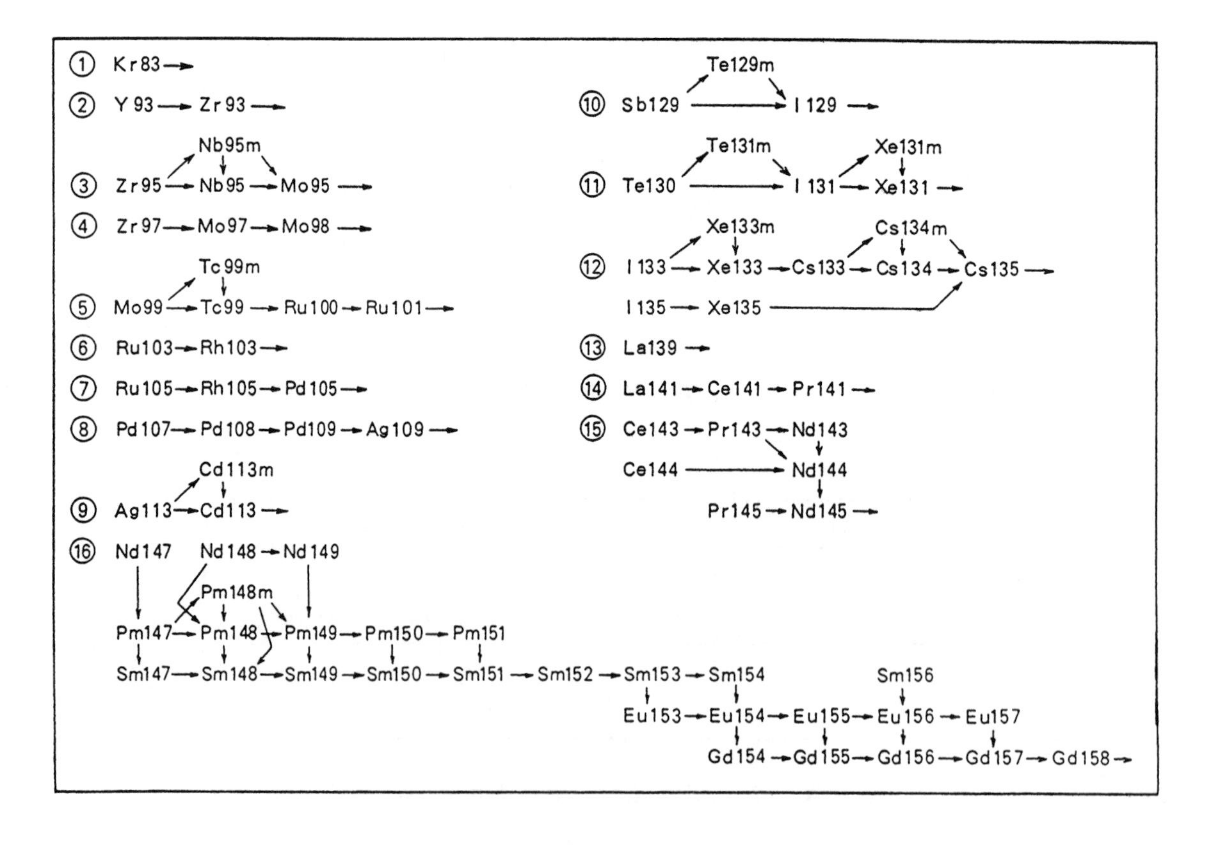

Figure B.5-2 **Fission product chains**

Appendix B.6

OECD/NEA benchmark on plutonium recycling in PWRs

D. C. LUTZ (IKE)

The cross-section base is the JEF-1 data file [1], which is processed [2] with the NJOY program system [3] into multigroup data. Three libraries are available in following energy ranges:

- Fast and epithermal range, 100 groups,
- The resolved resonance region, 8500 groups,
- The thermal range, 151 groups.

In these energy ranges the code CGM [4] performs one-dimensional cell-calculations with usually 5 zones (2 in the fuel, clad, 2 in the moderator) to get spectra for group collapsing to 45 (10 fast, 35 thermal) groups. The generated data sets are dependent of the cell definition in CGM, mainly in the groups containing resonances. They include the effects of self and mutual shielding and resonance overlapping for the U- and Pu-isotopes. This method of group cross-section generation has been used at IKE in principle for 20 years [5] with only few modifications to the number of groups and the size of the resonance region.

In standard calculations only 1 set of cross sections is produced for the whole life of the fuel, but in this case the cross sections have been recalculated 4 times during the irradiation.

The cell burnup calculations are carried out in RSYST 3 [6] for a cylindrical cell using a collision probability code. For 20 Actinides and 84 fission product isotopes the burnup equations are solved.

The fission yields are from JEF-1.

The fission spectrum has been recalculated after each burnup step using the actual fission rates of the U- and Pu-isotopes. The (n,2n)-reaction of U-238 and Pu-239 has been included in the actinide chain.

References

[1] J. L. Rowlands, N. Tubbs, "The Joint Evaluated File: A New Data Library for Reactor Calculations", Int. Conf. on Nuclear Data for Basic Applied Science, Santa Fe, 1985.

[2] Mattes, M.: IKE, Internal Report, 1989.

[3] R. E. Mac Farlane et al., "The NJOY Nuclear Data Processing System", LA-9309-M, 1985.

[4] M. Arshad, "Development and Validation of a Program System for Calculation of Spectra and Weighted Group Constants for Thermal and Epithermal Systems", Thesis University of Stuttgart (IKE 6-156), 1986.

[5] D. C. Lutz, "Einfluß der energetischen Abschirmung auf die epithermischen Gruppenquerschnitte bei einem DWR-Brennstab", Atomkernenergie 25, 85, 1975.

[6] R. Rühle, "RSYST An Integrated Modular System for Reactor and Shielding Calculations", USAEC Conf-730 414-12, 1973.

Appendix B.7

Plutonium recycling in PWRs

Monte Carlo calculations for the unburnt state

M. Mattes, W. Bernnat and S. Käfer (IKE)

Monte Carlo calculations have been performed for the fresh fuel using the continuous energy code MCNP-4.2. The cross section library based on JEF-2.2 was generated by NJOY-91.91 and contains data sets for the temperatures of 293.6 K, 600 K and 900 K. For hydrogen bound in H_2O a data set for a temperature of 573.6 K is available.

The calculations are done for

- T_f = 900 K,
- T_c = 600 K, and
- T_m = 573.6 K,

each for 995 000 histories.

The influence of the discrepancy in fuel temperature (33 K) on k-infinity could be in the range of 0.001.

For both cases calculations have been carried out also for a temperature of 293.6 K (335000/325000 histories) to be able to compare results using JEF-2.2 with results based on ENDF/B-V sent by ANL and performed by VIM using cross-sections at 300 K.

1 Benchmark A – Degraded plutonium – BOL — HOT

$k_{inf} = 1.13192 \pm 0.07$ % *

Table B.7-1.1 **Microscopic one-group cross-sections**

Nuclide	Absorption	uncertainty	Fission	uncertainty	Nu-bar
		Fuel			
^{235}U	1.6072E+01	0.16 %	1.1977E+01	0.16 %	2.4459
^{238}U	9.2939E-01	0.19 %	1.2153E-01	0.20 %	2.8080
^{238}Pu	9.2076E+00	0.23 %	1.8740E+00	0.16 %	3.0443
^{239}Pu	3.1774E+01	0.18 %	2.0326E+01	0.18 %	2.8797
^{240}Pu	2.0635E+01	0.24 %	6.8147E-01	0.14 %	3.0846
^{241}Pu	3.4975E+01	0.17 %	2.6666E+01	0.17 %	2.9400
^{242}Pu	7.2074E+00	0.44 %	5.1151E-01	0.15 %	3.1255
^{16}O	5.1595E-03	0.48 %			
		Cladding			
Zr	3.2878E-02	0.37 %			
		Moderator			
H	9.0128E-03	0.18 %			
^{16}O	4.8337E-03	0.48 %			
^{10}B	1.0432E+02	0.18 %			
^{11}B	1.9526E-04	0.47 %			

Table B.7-1.2 **Isotopic reaction rates**

Nuclide	Absorption	uncertainty	Fission	uncertainty
		Fuel		
^{235}U	2.6016E-02	0.11 %	1.9387E-02	0.11 %
^{238}U	2.0750E-01	0.14 %	2.7133E-02	0.16 %
^{238}Pu	1.1823E-02	0.18 %	2.4063E-03	0.11 %
^{239}Pu	3.6593E-01	0.13 %	2.3409E-01	0.13 %
^{240}Pu	1.8406E-01	0.19 %	6.0785E-03	0.09 %
^{241}Pu	1.3314E-01	0.18 %	1.0151E-01	0.10 %
^{242}Pu	4.5508E-02	0.39 %	3.2300E-03	0.10 %
^{16}O	2.6492E-03	0.43 %		
		Cladding		
Zr	5.5436E-03	0.32 %		
		Moderator		
H	9.3516E-03	0.14 %		
^{16}O	2.5078E-03	0.44 %		
^{10}B	8.2453E-03	0.14 %		
^{11}B	6.8896E-08	0.43 %		

Table B.7-1.3 **Flux spectrum by region (E_{max} = 20 MeV)**

Group	Elower (eV)	Flux	uncertainty
	Fuel		
1	8.21000E+05	3.5229E+00	0.12 %
2	5.53000E+03	4.5156E+00	0.07 %
3	6.25000E-01	2.9151E+00	0.07 %
4	1.00000E-05	2.4408E-01	0.19 %
Totals		1.1198E+01	0.05 %
	Cladding		
Group	Elower (eV)	Flux	uncertainty
1	8.21000E+05	1.1737E+00	0.12 %
2	5.53000E+03	1.5369E+00	0.07 %
3	6.25000E-01	1.0681E+00	0.07 %
4	1.00000E-05	1.2413E-01	0.19 %
Totals		3.8987E+00	0.05 %
	Moderator		
Group	Elower (eV)	Flux	uncertainty
1	8.21000E+05	6.3642E+00	0.12 %
2	5.53000E+03	8.4130E+00	0.07 %
3	6.25000E-01	6.1371E+00	0.07 %
4	1.00000E-05	8.3095E-01	0.17 %
Totals		2.1745E+01	0.04 %

* The MCNP output offers several slightly different eigenvalues resulting from different averaging methods. The eigenvalues listed in this Appendix are "simple average" values, the ones included in Tab.4A and 4B are "combined average" values being recommended in the User's Manual.

2 Benchmark B – Better plutonium – BOL — HOT

$k_{inf} = 1.18497 \pm 0.07$ % *

Table B.7-2.1 **Microscopic one-group cross-sections**

Nuclide	Absorption	uncertainty	Fission	uncertainty	Nu-bar
Fuel					
^{234}U	1.9890E+01	0.64 %	5.6624E-01	0.15 %	2.6344
^{235}U	2.3398E+01	0.16 %	1.8100E+01	0.16 %	2.4430
^{238}U	9.6419E-01	0.18 %	1.1629E-01	0.21 %	2.8101
^{238}Pu	1.4127E+01	0.21 %	1.9886E+00	0.15 %	3.0298
^{239}Pu	5.4075E+01	0.17 %	3.4762E+01	0.17 %	2.8747
^{240}Pu	4.1622E+01	0.27 %	6.6128E-01	0.14 %	3.0809
^{241}Pu	5.6056E+01	0.17 %	4.2394E+01	0.17 %	2.9369
^{242}Pu	2.2119E+01	0.65 %	4.9988E-01	0.15 %	3.1195
^{16}O	4.9919E-03	0.42 %			
Cladding					
Zr	3.4177E-02	0.34 %			
Moderator					
H	1.3434E-02	0.17 %			
^{16}O	4.6868E-03	0.48 %			
^{10}B	1.5545E+02	0.17 %			
^{11}B	2.6834E-04	0.36 %			

Table B.7-2.2 **Isotopic reaction rates**

Nuclide	Absorption	uncertainty	Fission	uncertainty
Fuel				
^{234}U	5.8675E-05	0.59 %	1.6704E-06	0.10 %
^{235}U	1.4439E-02	0.11 %	1.1169E-02	0.11 %
^{238}U	2.3441E-01	0.13 %	2.8272E-02	0.16 %
^{238}Pu	3.6891E-03	0.16 %	5.1932E-04	0.10 %
^{239}Pu	4.6091E-01	0.18 %	2.9630E-01	0.12 %
^{240}Pu	1.3772E-01	0.22 %	2.1882E-03	0.09 %
^{241}Pu	9.7977E-02	0.12 %	7.4096E-02	0.12 %
^{242}Pu	1.2623E-02	0.60 %	2.8529E-04	0.10 %
^{16}O	2.5774E-03	0.43 %		
Cladding				
Zr	6.1500E-03	0.29 %		
Moderator				
H	1.4881E-02	0.13 %		
^{16}O	2.5958E-03	0.44 %		
^{10}B	1.3116E-02	0.13 %		
^{11}B	1.0109E-07	0.32 %		

Table B.7-2.3 **Flux spectrum by region (E_{max} = 20 MeV)**

Group	Elower (eV)	Flux	uncertainty
Fuel			
1	8.21000E+05	3.6049E+00	0.12 %
2	5.53000E+03	4.5870E+00	0.07 %
3	6.25000E-01	3.2833E+00	0.07 %
4	1.00000E-05	5.0397E-01	0.16 %
Totals		1.1979E+01	0.05 %
Cladding			
1	8.21000E+05	1.2003E+00	0.12 %
2	5.53000E+03	1.5568E+00	0.07 %
3	6.25000E-01	1.1836E+00	0.06 %
4	1.00000E-05	2.2111E-01	0.16 %
Totals		4.1619E+00	0.05 %
Moderator			
1	8.21000E+05	6.5047E+00	0.12 %
2	5.53000E+03	8.5456E+00	0.07 %
3	6.25000E-01	6.7516E+00	0.05 %
4	1.00000E-05	1.4125E+00	0.15 %
Totals		2.3214E+01	0.04 %

* The MCNP output offers several slightly different eigenvalues resulting from different averaging methods. The eigenvalues listed in this Appendix are "simple average" values, the ones included in Tab.4A and 4B are "combined average" values being recommended in the User's Manual.

3 Benchmark A – Degraded plutonium – BOL — 293.6 K

$k_{inf} = 1.15862 \pm 0.11$ %

Table B.7-3.1 **Microscopic one-group cross-sections**

Nuclide	Absorption	uncertainty	Fission	uncertainty	Nu-bar
Fuel					
^{235}U	1.7157E+01	0.27 %	1.2887E+01	0.27 %	2.4452
^{238}U	8.7385E-01	0.32 %	1.2065E-01	0.34 %	2.8091
^{238}Pu	1.0054E+01	0.41 %	1.8902E+00	0.28 %	3.0422
^{239}Pu	3.1616E+01	0.29 %	2.0525E+01	0.29 %	2.8804
^{240}Pu	2.0882E+01	0.39 %	6.7642E-01	0.24 %	3.0851
^{241}Pu	3.6169E+01	0.28 %	2.7582E+01	0.27 %	2.9397
^{242}Pu	6.7528E+00	0.74 %	5.0786E-01	0.26 %	3.1260
^{16}O	5.1413E-03	0.81 %			
Cladding					
Zr	3.2936E-02	0.65 %			
Moderator					
H	1.0086E-02	0.34 %			
^{16}O	4.8206E-03	0.83 %			
^{10}B	1.1674E+02	0.34 %			
^{11}B	2.1656E-04	0.77 %			

Table B.7-3.2 **Isotopic reaction rates**

Nuclide	Absorption	uncertainty	Fission	uncertainty
Fuel				
^{235}U	2.7978E-02	0.19 %	2.0976E-02	0.19 %
^{238}U	1.9618E-01	0.24 %	2.7087E-02	0.26 %
^{238}Pu	1.2981E-02	0.33 %	2.4405E-03	0.20 %
^{239}Pu	3.6612E-01	0.21 %	2.3768E-01	0.21 %
^{240}Pu	1.8729E-01	0.31 %	6.0667E-03	0.16 %
^{241}Pu	1.3845E-01	0.20 %	1.0558E-01	0.19 %
^{242}Pu	4.2873E-02	0.66 %	3.2243E-03	0.18 %
^{16}O	2.6544E-03	0.73 %		
Cladding				
Zr	5.5829E-03	0.57 %		
Moderator				
H	1.0523E-02	0.26 %		
^{16}O	2.5146E-03	0.75 %		
^{10}B	9.2775E-03	0.26 %		
^{11}B	7.6830E-08	0.69 %		

Table B.7-3.3 **Flux spectrum by region (E_{max} = 20 MeV)**

Group	Elower (eV)	Flux	uncertainty
Fuel			
1	8.21000E+05	3.5125E+00	0.29 %
2	5.53000E+03	4.5095E+00	0.18 %
3	6.25000E-01	2.9743E+00	0.16 %
4	1.00000E-05	2.5339E-01	0.44 %
Totals		1.1250E+01	0.12 %
Cladding			
Group	Elower (eV)	Flux	uncertainty
1	8.21000E+05	1.1729E+00	0.30 %
2	5.53000E+03	1.5291E+00	0.18 %
3	6.25000E-01	1.0863E+00	0.16 %
4	1.00000E-05	1.2860E-01	0.46 %
Totals		3.9169E+00	0.12 %
Moderator			
Group	Elower (eV)	Flux	uncertainty
1	8.21000E+05	6.3613E+00	0.30 %
2	5.53000E+03	8.3993E+00	0.16 %
3	6.25000E-01	6.2320E+00	0.13 %
4	1.00000E-05	8.6475E-01	0.41 %
Totals		2.1858E+01	0.11 %

4 Benchmark B – Better plutonium – BOL — 293.6 K

$k_{inf} = 1.21816 \pm 0.11$ %

Table B.7-4.1 **Microscopic one-group cross-sections**

Nuclide	Absorption	uncertainty	Fission	uncertainty	Nu-bar
Fuel					
^{234}U	2.0405E+01	1.27 %	5.6593E-01	0.25 %	2.6341
^{235}U	2.6027E+01	0.28 %	2.0299E+01	0.29 %	2.4423
^{238}U	9.1057E-01	0.31 %	1.1602E-01	0.35 %	2.8110
^{238}Pu	1.6432E+01	0.38 %	2.0558E+00	0.29 %	3.0251
^{239}Pu	5.3738E+01	0.29 %	3.5204E+01	0.29 %	2.8760
^{240}Pu	4.1689E+01	0.46 %	6.5857E-01	0.25 %	3.0816
^{241}Pu	5.8769E+01	0.29 %	4.4379E+01	0.28 %	2.9367
^{242}Pu	2.0720E+01	1.23 %	4.9705E-01	0.26 %	3.1207
^{16}O	4.9847E-03	0.84 %			
Cladding					
Zr	3.4548E-02	0.62 %			
Moderator					
H	1.5463E-02	0.32 %			
^{16}O	4.6964E-03	0.85 %			
^{10}B	1.7892E+02	0.32 %			
^{11}B	3.0250E-04	0.60 %			

Table B.7-4.2 **Isotopic reaction rates**

Nuclide	Absorption	uncertainty	Fission	uncertainty
Fuel				
^{234}U	6.0652E-05	1.19 %	1.6822E-06	0.17 %
^{235}U	1.6184E-02	0.20 %	1.2622E-02	0.21 %
^{238}U	2.2306E-01	0.23 %	2.8420E-02	0.27 %
^{238}Pu	4.3238E-03	0.30 %	5.4094E-04	0.21 %
^{239}Pu	4.6154E-01	0.21 %	3.0235E-01	0.21 %
^{240}Pu	1.3900E-01	0.38 %	2.1958E-03	0.17 %
^{241}Pu	1.0350E-01	0.21 %	7.8160E-02	0.20 %
^{242}Pu	1.1916E-02	1.15 %	2.8583E-04	0.18 %
^{16}O	2.5932E-03	0.76 %		
Cladding				
Zr	6.2307E-03	0.54 %		
Moderator				
H	1.7174E-02	0.24 %		
^{16}O	2.6081E-03	0.77 %		
^{10}B	1.5137E-02	0.24 %		
^{11}B	1.1425E-07	0.52 %		

Table B.7-4.3 **Flux spectrum by region (E_{max} = 20 MeV)**

Group	Elower (eV)	Flux	uncertainty
Fuel			
1	8.21000E+05	3.6192E+00	0.21 %
2	5.53000E+03	4.5934E+00	0.13 %
3	6.25000E-01	3.3345E+00	0.12 %
4	1.00000E-05	5.2323E-01	0.26 %
Totals		1.2070E+01	0.08 %
Cladding			
1	8.21000E+05	1.2040E+00	0.22 %
2	5.53000E+03	1.5591E+00	0.13 %
3	6.25000E-01	1.1996E+00	0.11 %
4	1.00000E-05	2.2930E-01	0.27 %
Totals		4.1920E+00	0.08 %
Moderator			
1	8.21000E+05	6.5231E+00	0.21 %
2	5.53000E+03	8.5521E+00	0.11 %
3	6.25000E-01	6.8346E+00	0.09 %
4	1.00000E-05	1.4666E+00	0.25 %
Totals		2.3376E+01	0.08 %

Appendix B.8

Results of the benchmark for plutonium recycling in PWRs

H.Akie and H. Takano (JAERI)

Both benchmarks with degraded and better plutonium vectors were performed with the SRAC system.

The linear heat ratings of 183 W/cm for the degraded plutonium case and 172 W/cm for better Pu case were used in the SRAC burnup calculations, which were calculated from the power density of 30.3 W/g of initial heavy metal.

The following fission energies were used:

- U-235 193.7 + 8 = 201.7 MeV (= 3.232E-11 J in SRAC)
- U-238 197.0 + 8 = 205.0 MeV (= 3.284E-11 J)
- Pu-239 202.0 + 8 = 210.0 MeV (= 3.365E-11 J)
- Pu-241 204.4 + 8 = 212.4 MeV (= 3.403E-11 I)
- Am-242m 207.0 + 8 = 215.0 MeV (= 3.445E-11 1)

The requested description of the SRAC calculation.

- *Computer program and version*: modification version of SRAC [1].

- *Data libraries and original evaluated data file*: SRACLIB-JENDL3 processed from JENDL-3.1 [2] data file.

- *List of isotopes for resonance shielding calculation and method used*: The self-shielding factor table (f-table) interpolation method can be applied in the whole energy region. In the resolved resonance region (E < 961 eV in the SRAC system), PEACO [3] can treat both the self-shielding and the mutual resonance overlapping effects by the ultra-fine group method, which calculates the spectrum with the energy structure of lethargy width Δu = 2.5E-4 between 961 eV and 130 eV, and Δu = 5.0E-4 between 130 eV and 2.38 eV (2.38 eV is the thermal cut off energy in the calculations here).

The resonance shielding was not considered for Ru-105, I-135, Pm-151 and pseudo FP (and of course for H-1).

The PEACO method was used for

> U-233, U-234, U-235, U-236, U-238, Np-237, Np-239, Pu-238, Pu-239, Pu-240, Pu-241, Pu-242, Am-241, Am-242m, Am-242, Cm-244 and Ag-109.

The self-shielding effect in the thermal range (E<2.38 eV) was considered with f-table method for

> Th-230, U-233, U-235, Np-237, Pu-238, Pu-239, Pu-240, Pu-242, Am-241, Am-242m, Cm-243, Cm-245, Pm-148, Eu-153, Eu-155 and Eu-156.

For the other nuclides, the f-table method was used in the fast energy range.

- *Treatment of Xe build-up*: The build-up of Xe is treated accurately by the build-up and decay chain scheme as shown in Figure B.8-1.

- *(n,2n)-reaction treatment for the k-infinity calculation*: The (n,2n)-reaction rate is treated by subtracting from the total absorption rate.

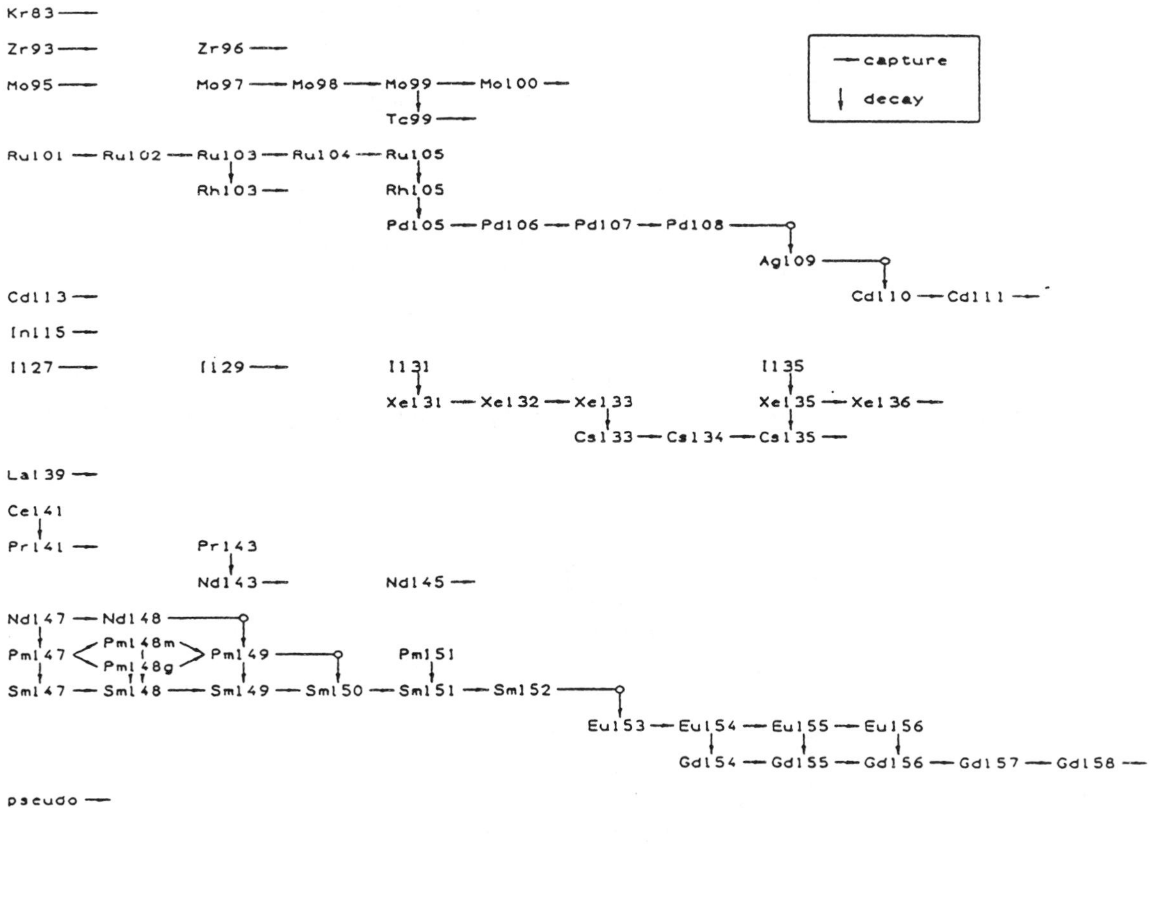

Kr83
Zr93 Zr96
capture
decay
Mo95 Mo97 Mo98 Mo99 Mo100
Tc99
Ru101 Ru102 Ru103 Ru104 Ru105
Rh103 Rh105
Pd105 Pd106 Pd107 Pd108
Ag109
Cd113 Cd110 Cd111
In115
I127 I129 I131 I135
Xe131 Xe132 Xe133 Xe135 Xe136
Cs133 Cs134 Cs135
La139
Ce141
Pr141 Pr143
Nd143 Nd145
Nd147 Nd148
Pm147 Pm148m Pm148g Pm149 Pm151
Sm147 Sm148 Sm149 Sm150 Sm151 Sm152
Eu153 Eu154 Eu155 Eu156
Gd154 Gd155 Gd156 Gd157 Gd158
pseudo

Figure B.8-1 ***Burnup chain scheme of FP nuclides in the SRAC system***

Energy range to evaluate (n,2n)-cross-sections and the effect on isotope production

In the SRAC system, the upper limit energy to consider all the reactions is 10 MeV, while a part of (n,2n)-reaction is included in the upper range of 10 MeV. It means that SRAC underestimates the (n,2n)-reaction and therefore the production of minor isotopes such as Np-237, which is mainly produced through the (n,2n)-reaction of U-238. For this reason, the effect of the energy range to treat (n,2n)-reaction was studied. Table B.8-1 shows the (n,2n)-cross-sections of U-238 in the degraded plutonium case cell evaluated for different energy ranges with the continuous-energy Monte Carlo code MVP. For 50 MWd/kg the Monte Carlo calculation was made with the simulated fuel composition assumed from the SRAC result. It can be seen that about 15% of the (n,2n)-reaction takes place over 10 MeV in this cell. Taking into account the difference, the SRAC cell burnup calculations were made for the degraded plutonium cell. Figure B.8-2 compares the number densities of Np-237. The Np-237 density becomes larger by about 9% at 50 MWd/kg when the contribution of the (n,2n)-reaction is taken into account up to 20 MeV.

Fuel Temp.	Energy Range	$\sigma(n,2n)$ ($\times 10^{-3}$ barn) 0GWd/t	50GWd/t
300K	≤20MeV	5.89±2.4%	5.85±2.5%
	≤10MeV	5.02±2.5%	4.98±2.4%
		(1.174)	(1.175)
900K	≤20MeV	5.71±2.5%	5.62±2.1%
	≤10MeV	4.87±2.6%	4.75±2.2%
		(1.172)	(1.183)

( ) : ratios of ≤20MeV/≤10MeV cases

Table B.8-1

U-238 (n,2n)-cross-sections calculated with MVP code for different energy ranges (degraded Pu cell)

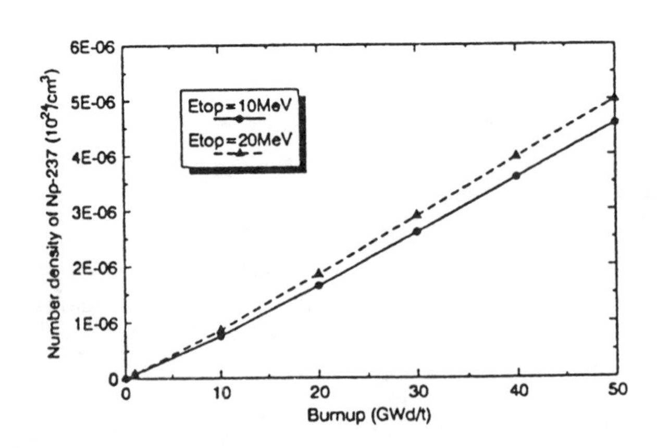

Figure B.8-2

Density of Np-237 calculated with U-238 $\sigma(n,2n)_S$ evaluated for different upper limit energy(Etop) (degraded plutonium cell)

References

[1] K. Tsuchihashi and Y. Ishiguro, "Revised SRAC Code System", JAERI 1302 (1986).

[2] K. Shibata and T. Nakagawa, "Japanese Evaluated Nuclear Data Library, Version-3", JAERI 1319 / NEANDC-J-150-U / INDC-JPN-137-L (1990).

[3] Y. Ishiguro, JAERI-M 5527 (1974).

Appendix B.9

MOX-PWR benchmark:

PSI Results from BOXER

J. M. Paratte (PSI)

Computer codes

The results given below were obtained using the ETOBOX and BOXER codes of the ELCOS light water neutronics code package [1] [2]. For some of the nuclides which are not included in the library for BOXER the densities were calculated with the code SELECT. All these codes were developed at PSI.

ETOBOX processes cross section data in ENDF/B format and produces a cross-section library for BOXER. BOXER performs cell, two-dimensional transport, and depletion calculations. SELECT is a depletion code based on one-group cross-sections which can handle a large number of nuclides.

The cross-section library produced by ETOBOX contains microscopic neutron cross-sections collapsed to 70 groups. The group structure is the 69 group WIMS structure with an extra group between 10 and 15 MeV. However, the upper boundary of the thermal energy range is 1.3 eV instead of 4 eV. P0 and P1 scattering matrices (P2 transport corrected) are given for most nuclides. The weighting spectrum is a spectrum calculated in many microgroups for a typical LWR-cell in the fast range, a l/E spectrum at intermediate energies, and a modified Maxwellian spectrum in the thermal range. In the fast range (E > 907 eV) the resonance cross-sections (both resolved and unresolved resonances) are Doppler-broadened and collapsed to groups for three temperatures and 4 values of the dilution cross-section. In the resonance range between 1.3 eV and 907 eV (important low-energy resonances of plutonium isotopes are included) pointwise lists of Doppler-broadened cross-sections are produced for three to seven temperatures (depending on the nuclide). For the unresolved resonances these lists are produced for four dilutions. The spacing of the points depends on the variation of the cross-sections with lethargy, so that the cross-section values between the points can be accurately reconstructed by interpolation. The minimum spacing of the points is 1.0E-5 lethargy units. Typical numbers of energy points for actinides are 7000 to 8000 between 1.3 and 907 eV. The thermal scattering matrices for most nuclides are calculated using the free gas model. For the moderator nuclides and especially for hydrogen in water the $S(\alpha, \beta)$ matrices given in the basic cross-section files are used.

In BOXER the resonance cross sections are self-shielded by a two region collision probability calculation in about 8000 lethargy points between 1.3 and 907 eV.

The fluxes in fine groups and in each zone are calculated by means of an integral transport method in cylindrical geometry. The fission source is assumed to be flat over all zones containing fissile nuclides. The scattering source in each zone can be flat or represented by a polynomial of the radius. In the epithermal range (above 1.3 eV) P1 corrected isotropic scattering is used. In the thermal range P1 anisotropy can be taken into account. The cells are calculated with white boundary conditions or with the outgoing partial current from a previously calculated cell as a fixed source at the periphery. The fundamental mode spectrum (i.e., for k-effective = 1) is determined by a B1 leakage calculation for the homogenised cell in 70 groups with an iterative search for the critical buckling [1]. Depletion calculations are performed using reaction rates collapsed to one group by weighting with the multigroup fluxes from the cell calculations (in the case one cell only is depleted). The time dependence of the nuclide densities is described by Taylor series with a given number of terms. The densities of nuclides with high destruction rates are calculated analytically with an exponential approach to their asymptotic densities. An iterative correction adjusts the flux within the time step in order to keep the power constant. The effect of the changing spectrum on the reaction rates is taken into account by a predictor-corrector method and by density-dependent one-group cross-sections within the time step for Pu-239 and Pu-240 (approximated by a rational function). A time step can be divided into several micro-steps without recalculating the reaction rates in order to improve the numerical accuracy of the depletion calculation.

Data library

The BOXLIB cross-section library for BOXER used in the present calculations contains cross-sections for 29 actinide nuclides (from U-234 through Cm-248), 55 fission products considered explicitly, and two pseudo fission products. The 55 fission products were chosen based on their contribution to the total fission product neutron absorption in LWR configurations; in addition six gadolinium isotopes are included for burnable poison calculations. For some fission products which contribute little to the absorption the resonance cross-sections are given for infinite dilution only.

The source of cross-section data for all nuclides is JEF-1, except for Gd-155, whose cross-sections are taken from JENDL-2. The fission product yields are taken from JEF-2 for thermal fission. The half lives for the radioactive nuclides were taken from [3] and [4].

The fission energies from JEF-1 were increased to take into account the capture energy of the excess neutrons. The increment varies between 8 and 15 MeV depending on the fissile nuclide considered.

The methods used in the cross-section processing are described above.

Results

The results for the two fuel cells of the benchmark were provided separately for each cell according to the definition of the benchmark for the beginning of life (burnup = 0) and 4 burnup points. The infinite multiplication factors k-infinity were given for each point of the burnup calculation (total 26 points). Most of the results were calculated with BOXER with the following exceptions:

- The nuclides Zr-95, Ru-106, Pd-106, Cd-113, In-115 and Cs-137, are not on the BOXLIB library, so their cross-sections cannot be correctly calculated through a spectrum calculation of the cell. The cross-sections used in SELECT have then to be considered as fairly rough

[1] This option can be dropped giving a value 0 to the input cell buckling as in the present calculations of the MOX cells.

approximations. Therefore the SELECT results provided are affected by a certain unknown inaccuracy. It is believed that the densities for Zr-95, Ru-106, Pd-106 and Cs-137 are realistic because the self-shielding of their cross-sections is not very important. On the other hand the densities for Cd-113 and In-115 may be in error by as much as a factor of 10, so the SELECT results are not provided.

- For Mo-95 the BOXER results are not correct because the precursors Zr-95 and Nb-95 were not taken into account. For this reason the Mo-95 density calculated by BOXER was replaced by the SELECT one. In this case the cross-section of Mo-95 used in SELECT is taken from the BOXER results.

References

[1] J. M. Paratte, K. Foskolos, P. Grimm and C. Maeder, "Das PSI-Codesystem ELCOS zur stationären Berechnung von Leichtwasserreaktoren", Proceedings of Jahrestagung Kerntechnik, Travemunde, p. 59 (1988).

[2] P. Grimm and J. M. Paratte, "Validation of the EIR LWR Calculation Methods for Criticality Assessment of Storage Pools", EIR Report 603 (1986).

[3] W. Seelmann-Eggebert et al., "Karlsruhe Chart of the Nuclides", 5th edition, Kernforschungszentrum Karlsruhe (1981).

[4] R. C. Weast, M. J. Astle, "CRC Handbook of Chemistry and Physics", 60th edition, CRC Press, Boca Raton (Florida, 1981).

Appendix B.10

PSI results generated by CASMO

Frank Holzgrewe (PSI)

Computer code

The results given here were generated using the CASMO Code Version 4.7 [1] [2]. PSI acquired a licence for the executable of this program from Studsvik Of America, the CASMO source program is not available.

CASMO is a multigroup two-dimensional transport code for burnup calculations on BWR and PWR assemblies or simple pin cells. The code handles a geometry consisting of cylindrical fuel rods of varying compositions, in a square pitch array.

The resonance region is defined to lie between 4 eV and 9118 eV. Resonance absorption above 9118 eV is regarded as being unshielded. The 1-eV resonance in Pu-240 and the 0.3-eV resonance in Pu-239 are considered to be adequately covered by the concentration of thermal groups around these resonances and are consequently excluded from the special resonance treatment. Four nuclides, U-235, U-236, U-238 and Pu-239 are treated as resonance absorbers.

The effective absorption and fission cross-sections in the resonance energy-region for important resonance absorbers are calculated using an equivalence theorem. This theorem relates tabulated effective resonance integrals for each resonance absorber in each resonance group to the particular heterogeneous problem. The effective resonance integrals are obtained by interpolation from tables of homogeneous resonance integrals in the data library. The homogeneous resonance integrals are tabulated with potential cross-section σ_p and temperature T as parameters and the interpolation is based on a $\sqrt{\sigma_p}$ and $\sqrt{T}$ dependence. A first order correction for the interaction associated with the presence of several nuclides in the same material is included. The basic principles for the resonance treatment are similar to those in the code WIMS [3]. The calculation of the fuel-to-fuel collision probability in an infinite uniform lattice partly follows a description given by [4].

The isotopic depletion as a function of irradiation is calculated for each fuel pin and for each region containing a burnable absorber. The burnup chains, with the isotopes linked through absorption and decay, are linearized and 24 separate fission products, 2 pseudo fission products and 17 heavy nuclides are treated. A predictor-corrector approach is used for the burnup calculation. For each burnup step the depletion is calculated twice, first using the spectra at the start of the step and then, after a new spectrum calculation, using the spectra at the end of the step. Average number densities from these two calculations are used as start values for the next burnup step.

CASMO has an option to estimate the equilibrium xenon number density at zero burnup. This option was not used for the benchmark calculations. The xenon-concentration was put equal to zero for the first burnup step (0 MWd/kg). Xenon is thereafter built in through the I-Xe chain.

Data library

The neutron data library is based on data from ENDF/B-IV. It contains-cross sections for 93 materials, most of which are individual nuclides. A few materials are either elements of natural composition or mixtures of elements.

Microscopic cross-sections are tabulated in 40 energy groups, covering the energy range from 0-10 MeV. This group structure, shown in the tables for the normalised neutron spectrum, is a condensation from the 69-group WIMS structure with an additional boundary at 1.855 eV.

The library contains absorption, fission, ν•fission, transport and P0 scattering cross-sections (P1 scattering cross sections are also included for hydrogen, deuterium and oxygen). Data are tabulated as function of temperature. For U-235, U-236, U-238 and Pu-239 shielded resonance integrals versus potential background cross-sections and temperature are tabulated. (n,2n)-cross-sections are not listed in the library but (n,2n)-reactions are taken into account by reducing the absorption cross section so that:

$$\sigma_{a,g}^{lib} = \sigma_{c,g} + \sigma_{f,g} - \sigma_{(n,2n),g}$$

Nuclides are identified by an ID number, which in general is chosen so that the first digits are equal to the atomic number of the nuclide and the last three digits show the isotope number. Fission products which are not separately treated are lumped together into two pseudo nuclides, one non-saturating (ID = 401) and one slowly saturating (ID = 402)

Results

The results for the two fuel cells of the benchmark were given for each cell according to the benchmark specifications. The burnup calculation was made for a total of 26 steps. All results are given for the 5S burnup points, 0, 10, 33, 42 and 50 MWd/kg and only the infinite multiplication factor is given for all 26 steps. All nuclides are identified by an ID number as described above, the ID number 61248 stands for Pm-148m.

CASMO does not treat all the fission products listed in the benchmark specifications. Those which were not considered are:

> Zr-95, Mo-95, Tc-99, Ru-106, Pd-106, Cd-113, In-115, I-129, Cs-137, Ce-144, Nd-144, Nd-148, Gd-156, Gd-157

CASMO takes only two pseudo fission products into consideration instead of the required four, one for non-saturating (ID = 401) and one for slowly saturating (ID = 402) nuclides. The neutron cross-section library for CASMO contains only data for Zr-2 and Zr-4 and therefore the cladding could not be specified as natural Zirconium.

References

[1] B. H. Forssen, M. Edenius, "CASMO-3, A Fuel Assembly Burnup Program", User's Manual - Version 4.7, Studsvik Of America NFA-8913, Rev.2.

[2] B. H. Forssen, H. Haggblom, M. Edenius, "CASMO-3, A Fuel Assembly Burnup Program", Methodology - Version 4.4, Studsvik Of America NFA-89/2.

[3] J. R. Askew, F. J. Fayers, P. B. Kemshell, "A General Description of the Lattice Code WIMS" JBNES 5 (1966), p. 564.

[4] Stammler et al, "Equivalence Relations for Resonance Integral Calculations" Journal of Nuclear Energy 27 (1973), p. 885.

Appendix B.11

NEA MOX pincell benchmark

Kim Ekberg (Studsvik)

Computer code used: CASMO-4.

Method used: CASMO-4 is a multigroup transport code for cross-section production for LWR. The production is co-ordinated with the requirements of the reactor analysis code SIMULATE-3, but the cross-sections can also be used in other codes. CASMO-4 and its predecessor CASMO-3 can handle all known LWR fuel designs from the commercial fuel vendors. The calculation was done with the CASMO-4 default burnup steps, with the addition of the specified steps.

Library used: In addition to the standard library for CASMO-4 several libraries under development do exist. In this study a 70-group library has been used, based on JEF 2.2. At present this library represents 30 fission products separately. The remaining fission products are represented by two pseudo fission products: one non-saturating and one slowly saturating. Some of the fission products listed in the benchmark specification are at present represented by the pseudo fission products.

Resonance treatment: An equivalence theorem is used to relate the heterogeneous problem to an equivalent homogeneous problem. The effective resonance integrals are obtained by interpolation from tables of homogeneous resonance integrals in the data library. The homogeneous resonance integrals are tabulated with potential cross-section and temperature as parameters and the interpolation is based on a $\sqrt{\sigma_p}$ and $\sqrt{T}$ dependence. A first order correction for the interaction associated with the presence of several nuclides in the same material is included.

The following nuclides are treated as resonance absorbers:

> U-235, U-236, U-238, Pu-239, Pu-240, Pu-241, Pu-242, Am-241, Am-242m, Er-167, Gd-155, Gd-156, Gd-157, Gd-158.

Results: Tables were given for the two fuel cells of the benchmark for the following parameters: Number densities, k-infinity, absorption reaction rates, ν•fission reaction rates, and ν values (total). Absorption and ν•fission rates are normalised to one fission neutron per second. Results are given for exposure values 0, 0.15, 10, 33, 42 and 50 MWd/kg. In addition is given the CASMO-4 summary table showing k-infinity, M^2, and wt % of U-235, fissile Pu and total Pu for the depletion steps used in the calculation.

Reference

M. Edenius, K. Ekberg, B. H. Forssen, D. Knott, "CASMO-4, A Fuel Assembly Burnup Program", User's Manual, STUDSVIK/SOA-93/1 (Restricted Distribution).

Appendix C

Plots of absorption rates, fission rates and spectra

A-ar

Benchmark A: Absorption Rate (Total normalized to 1)

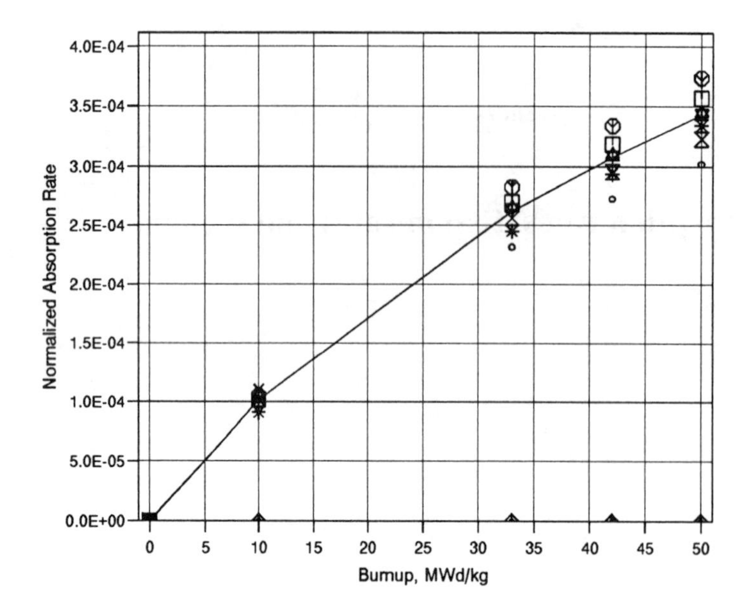

Benchmark A: Absorption Rate (Total normalized to 1)

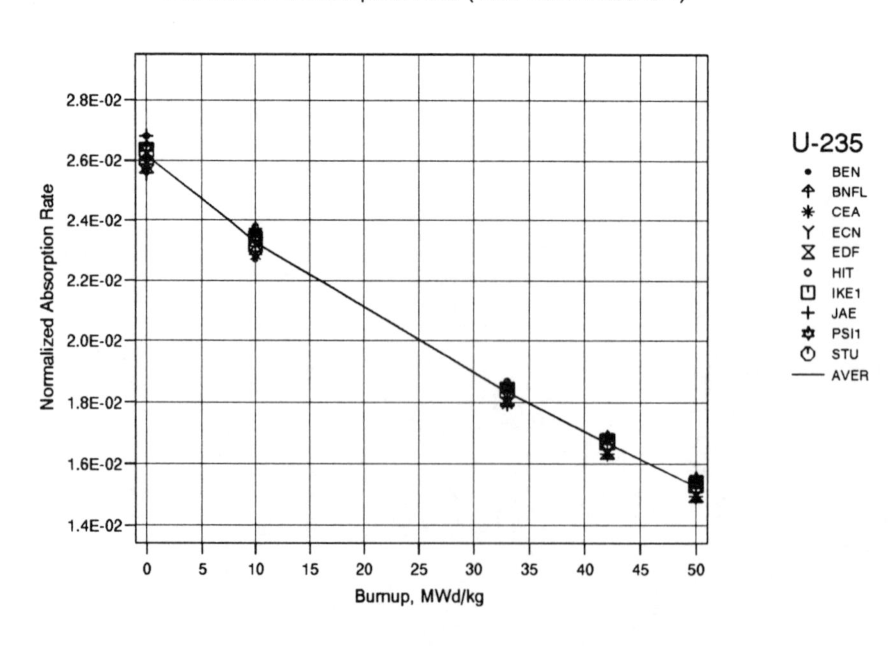

Benchmark A: Absorption Rate (Total normalized to 1)

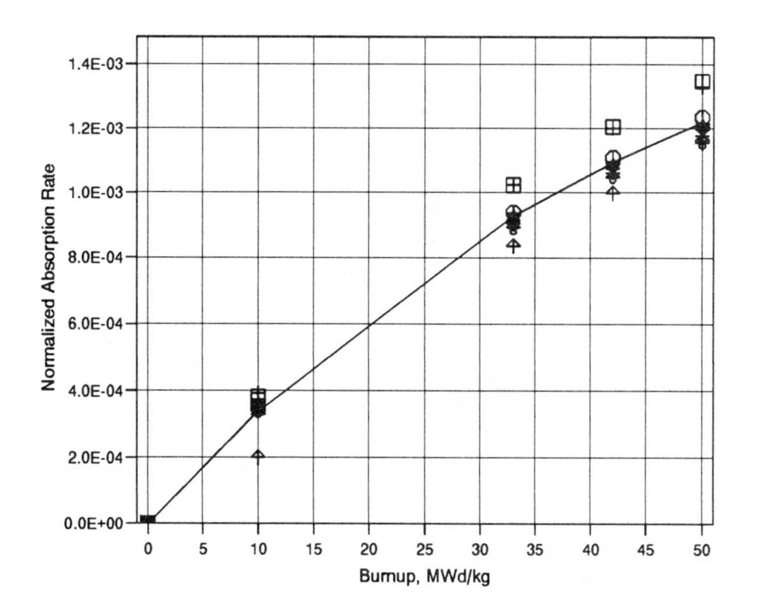

U-236

- BEN
- BNFL
- CEA
- ECN
- EDF
- HIT
- IKE1
- JAE
- PSI1
- STU
- AVER

Benchmark A: Absorption Rate (Total normalized to 1)

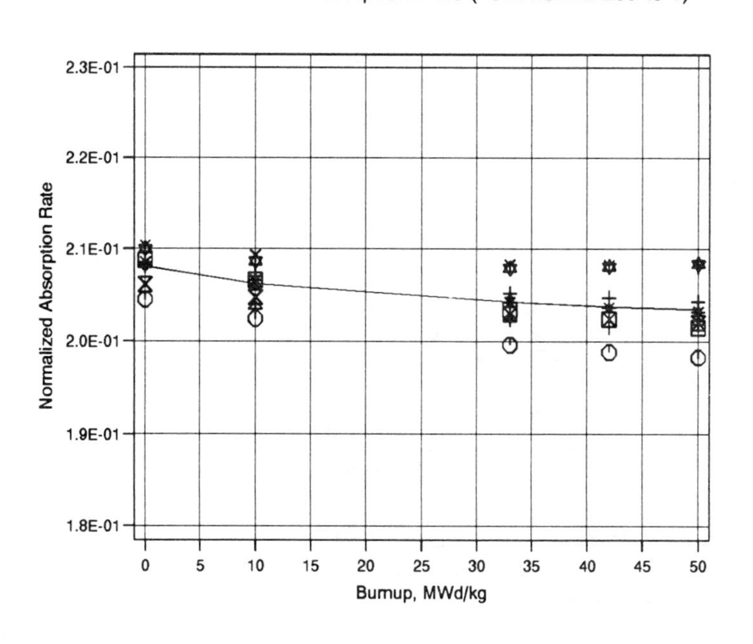

U-238

A-ar

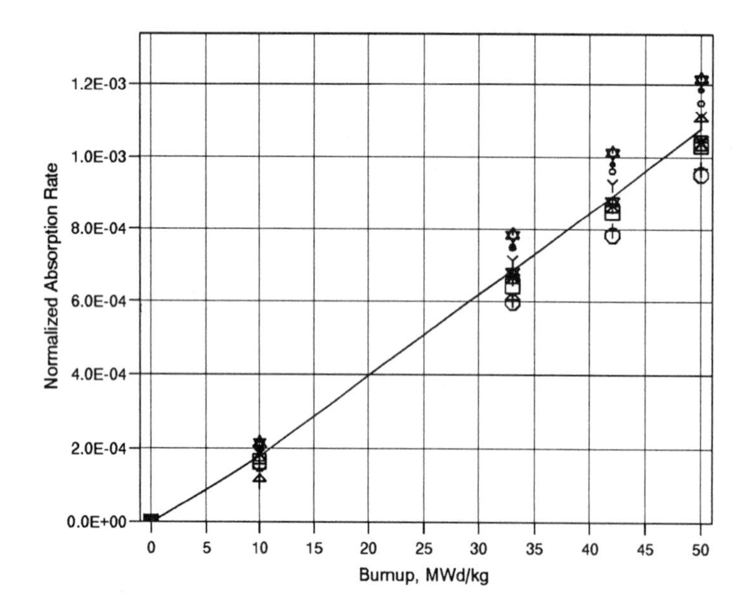
Benchmark A: Absorption Rate (Total normalized to 1)
Normalized Absorption Rate
Burnup, MWd/kg
Np-237
BEN
BNFL
CEA
ECN
EDF
HIT
IKE1
JAE
PSI1
STU
AVER

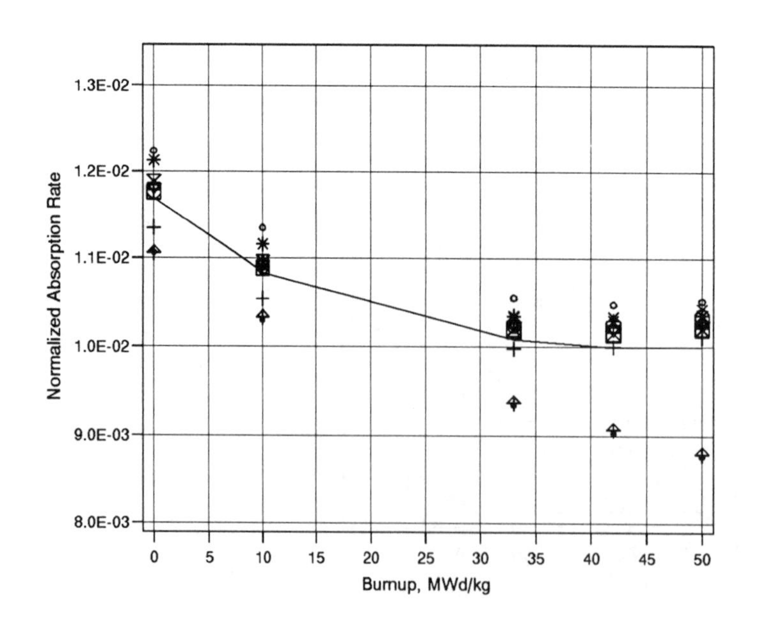
Benchmark A: Absorption Rate (Total normalized to 1)
Normalized Absorption Rate
Burnup, MWd/kg
Pu-238
BEN
BNFL
CEA
ECN
EDF
HIT
IKE1
JAE
PSI1
STU
AVER

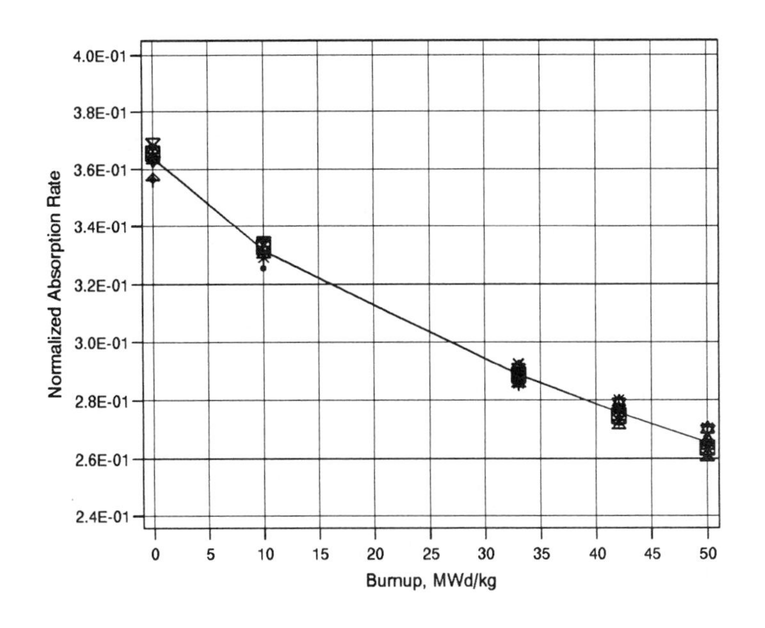
Benchmark A: Absorption Rate (Total normalized to 1)
4.0E-01
3.8E-01
3.6E-01
3.4E-01
3.2E-01
3.0E-01
2.8E-01
2.6E-01
2.4E-01
Normalized Absorption Rate
0
5
10
15
20
25
30
35
40
45
50
Burnup, MWd/kg
Pu-239
BEN
BNFL
CEA
ECN
EDF
HIT
IKE1
JAE
PSI1
STU
AVER

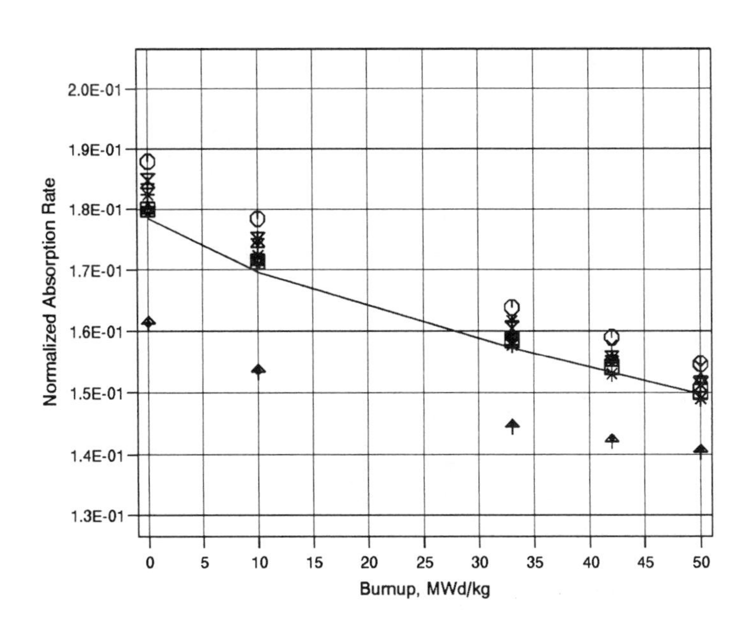
Benchmark A: Absorption Rate (Total normalized to 1)
2.0E-01
1.9E-01
1.8E-01
1.7E-01
1.6E-01
1.5E-01
1.4E-01
1.3E-01
Normalized Absorption Rate
0
5
10
15
20
25
30
35
40
45
50
Burnup, MWd/kg
Pu-240
BEN
BNFL
CEA
ECN
EDF
HIT
IKE1
JAE
PSI1
STU
AVER

A-ar

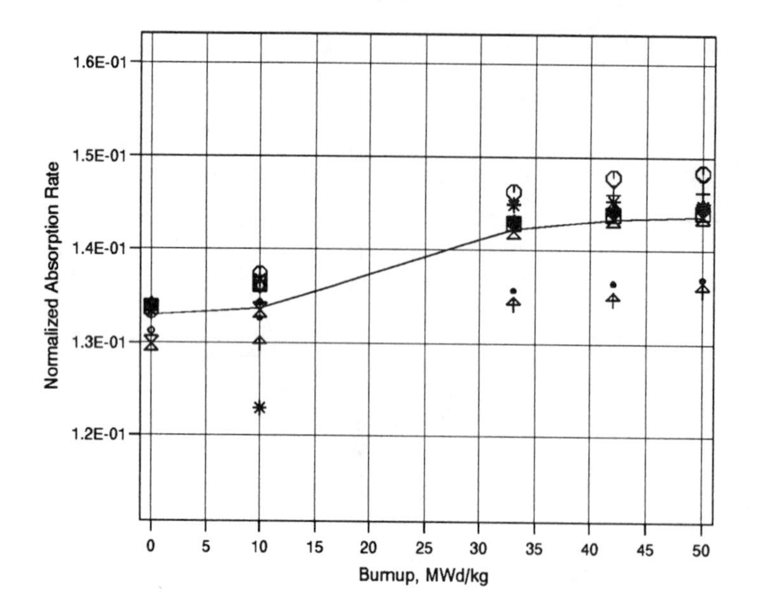
Benchmark A: Absorption Rate (Total normalized to 1)
1.6E-01
1.5E-01
1.4E-01
1.3E-01
1.2E-01
Normalized Absorption Rate
0
5
10
15
20
25
30
35
40
45
50
Burnup, MWd/kg
Pu-241
BEN
BNFL
CEA
ECN
EDF
HIT
IKE1
JAE
PSI1
STU
AVER

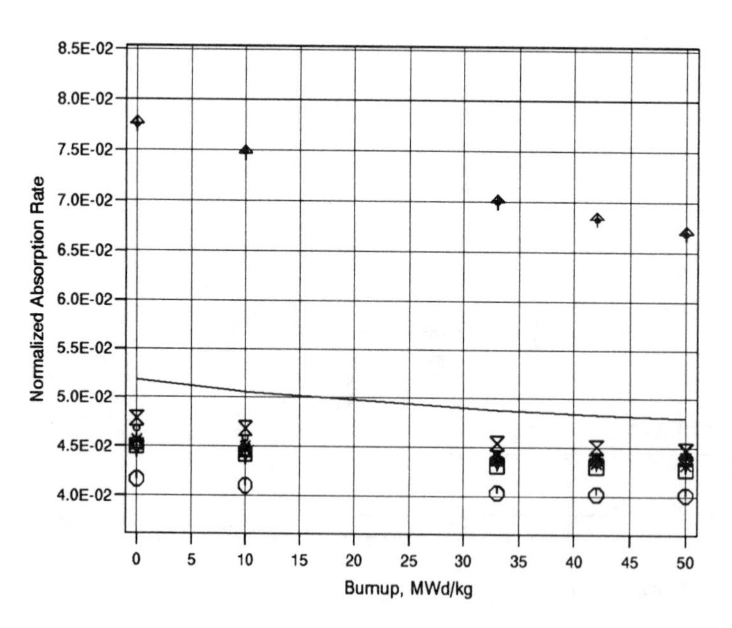
Benchmark A: Absorption Rate (Total normalized to 1)
8.5E-02
8.0E-02
7.5E-02
7.0E-02
6.5E-02
6.0E-02
5.5E-02
5.0E-02
4.5E-02
4.0E-02
Normalized Absorption Rate
0
5
10
15
20
25
30
35
40
45
50
Burnup, MWd/kg
Pu-242
BEN
BNFL
CEA
ECN
EDF
HIT
IKE1
JAE
PSI1
STU
AVER

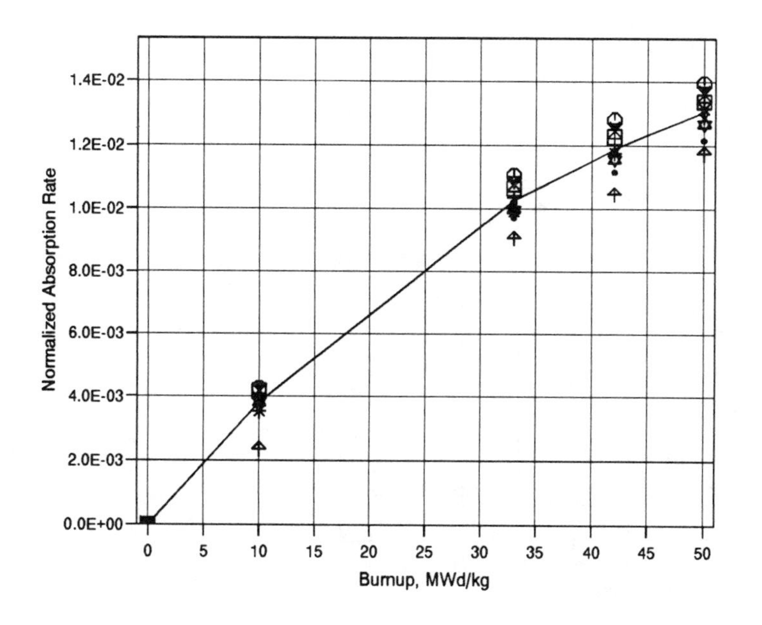
Benchmark A: Absorption Rate (Total normalized to 1)
Normalized Absorption Rate
1.4E-02
1.2E-02
1.0E-02
8.0E-03
6.0E-03
4.0E-03
2.0E-03
0.0E+00
0
5
10
15
20
25
30
35
40
45
50
Burnup, MWd/kg
Am-241
BEN
BNFL
CEA
ECN
EDF
HIT
IKE1
JAE
PSI1
STU
AVER

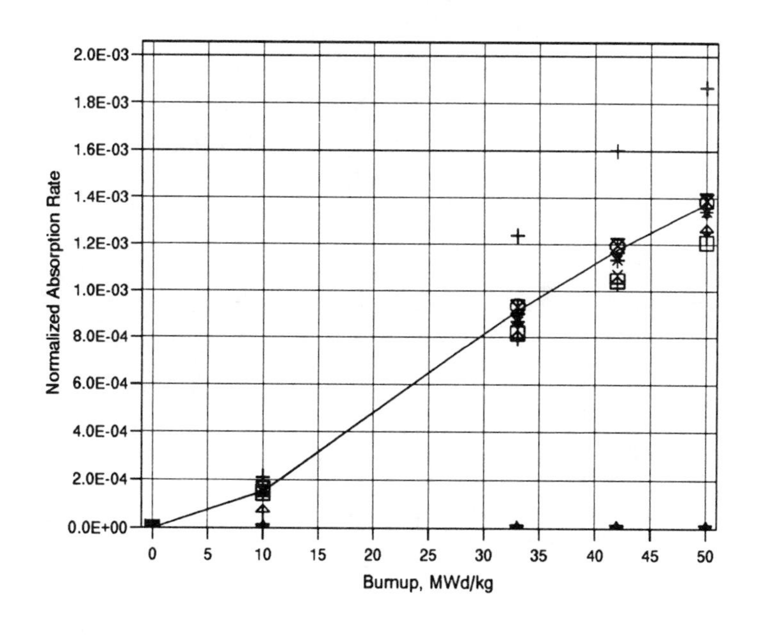
Benchmark A: Absorption Rate (Total normalized to 1)
Normalized Absorption Rate
2.0E-03
1.8E-03
1.6E-03
1.4E-03
1.2E-03
1.0E-03
8.0E-04
6.0E-04
4.0E-04
2.0E-04
0.0E+00
0
5
10
15
20
25
30
35
40
45
50
Burnup, MWd/kg
Am-242m
BEN
BNFL
CEA
ECN
EDF
HIT
IKE1
JAE
PSI1
STU
AVER

A-ar

Benchmark A: Absorption Rate (Total normalized to 1)

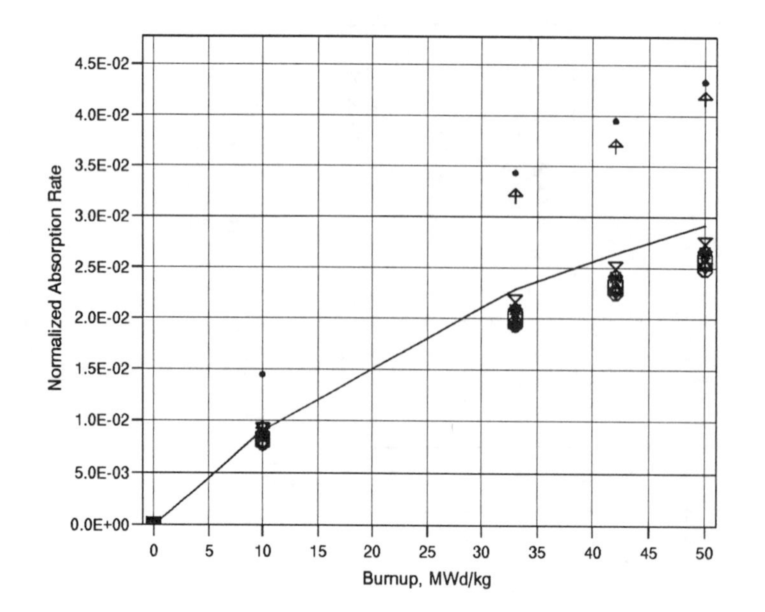

Am-243
- BEN
- BNFL
- CEA
- ECN
- EDF
- HIT
- IKE1
- JAE
- PSI1
- STU
- AVER

Benchmark A: Absorption Rate (Total normalized to 1)

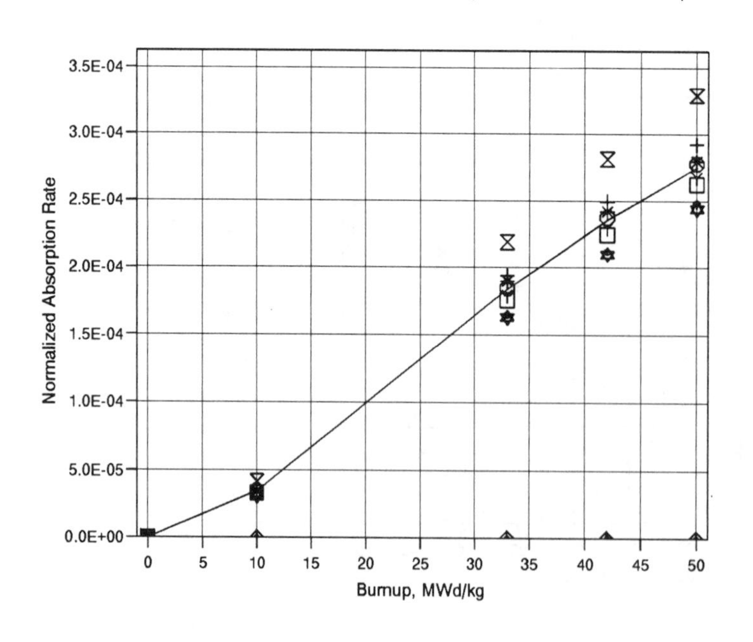

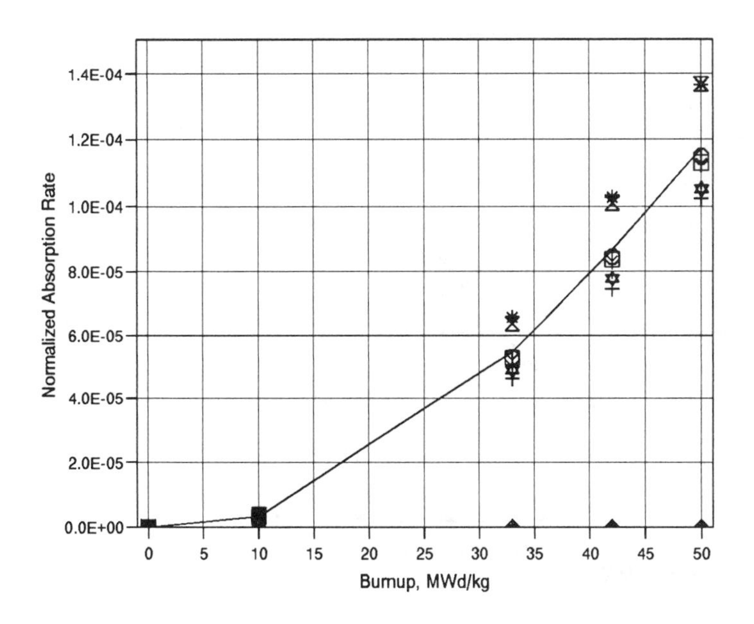
Benchmark A: Absorption Rate (Total normalized to 1)
Normalized Absorption Rate
Burnup, MWd/kg
Cm-243
BEN
BNFL
CEA
ECN
EDF
HIT
IKE1
JAE
PSI1
STU
AVER

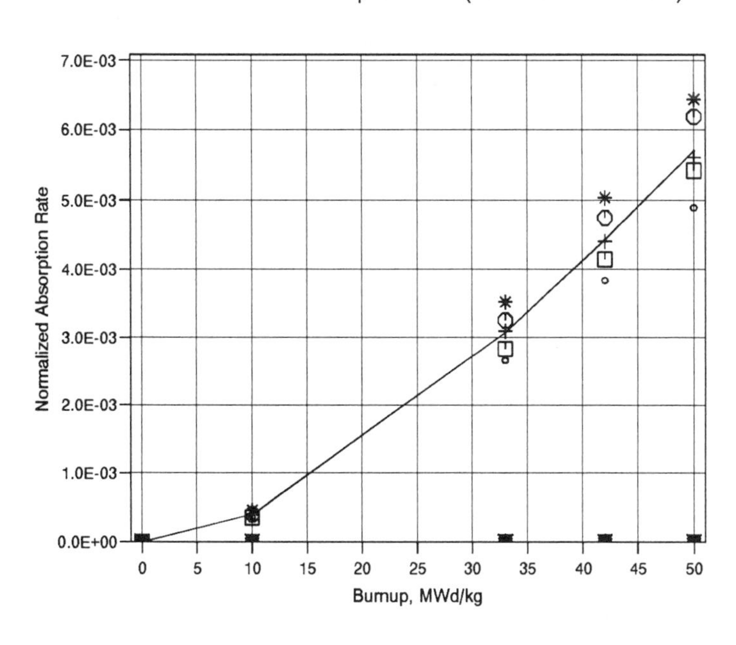
Benchmark A: Absorption Rate (Total normalized to 1)
Normalized Absorption Rate
Burnup, MWd/kg
Cm-244
BEN
BNFL
CEA
ECN
EDF
HIT
IKE1
JAE
PSI1
STU
AVER

A-ar

Benchmark A: Absorption Rate (Total normalized to 1)

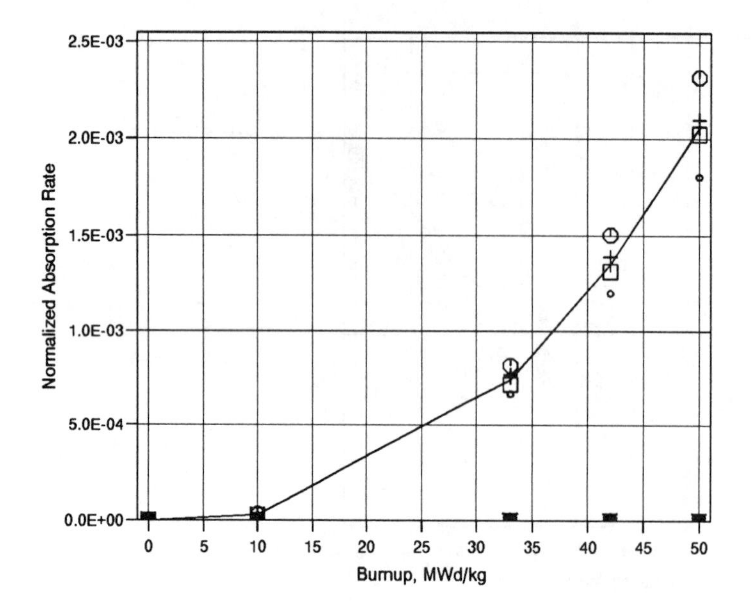

A-fr

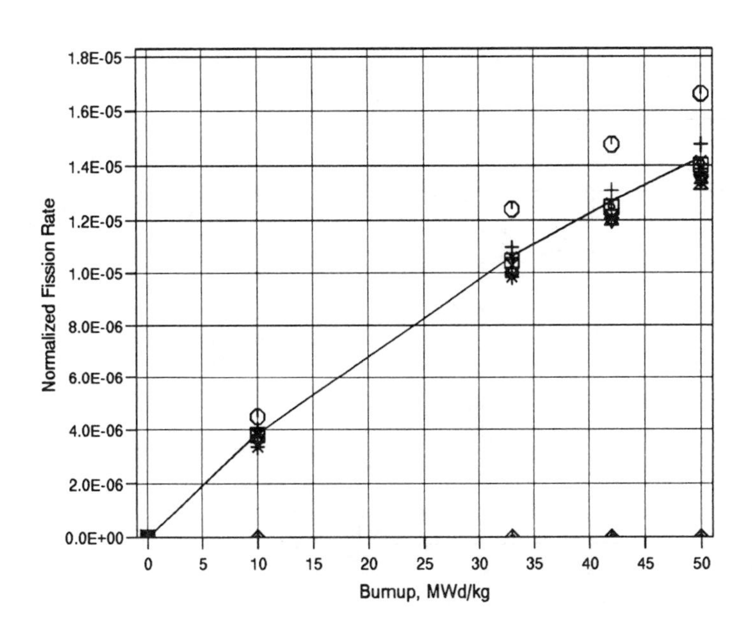
Benchmark A: Fission Rate
Normalized Fission Rate
1.8E-05
1.6E-05
1.4E-05
1.2E-05
1.0E-05
8.0E-06
6.0E-06
4.0E-06
2.0E-06
0.0E+00
0
5
10
15
20
25
30
35
40
45
50
Burnup, MWd/kg
U-234
BEN
BNFL
CEA
ECN
EDF
HIT
IKE1
JAE
PSI1
STU
AVER

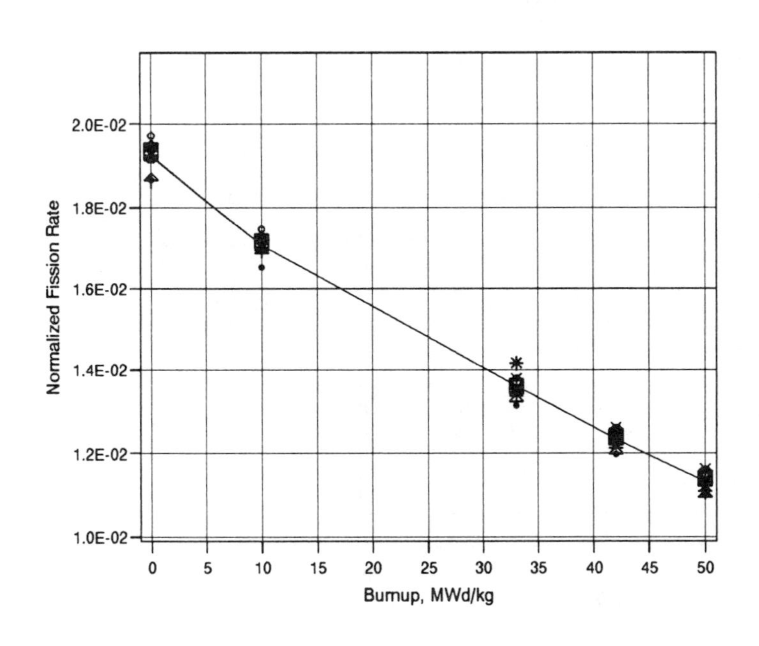
Benchmark A: Fission Rate
Normalized Fission Rate
2.0E-02
1.8E-02
1.6E-02
1.4E-02
1.2E-02
1.0E-02
0
5
10
15
20
25
30
35
40
45
50
Burnup, MWd/kg
U-235
BEN
BNFL
CEA
ECN
EDF
HIT
IKE1
JAE
PSI1
STU
AVER

A-fr

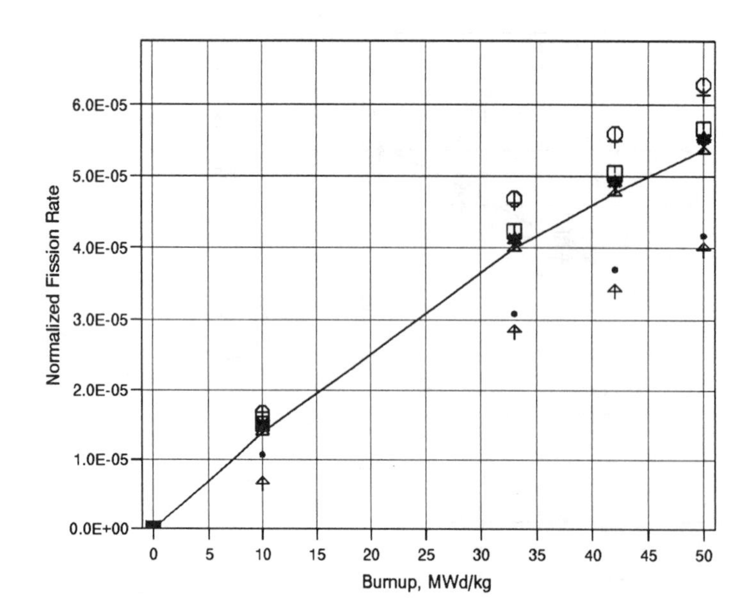
Benchmark A: Fission Rate
6.0E-05
5.0E-05
4.0E-05
3.0E-05
2.0E-05
1.0E-05
0.0E+00
Normalized Fission Rate
0
5
10
15
20
25
30
35
40
45
50
Burnup, MWd/kg
U-236
BEN
BNFL
CEA
ECN
EDF
HIT
IKE1
JAE
PSI1
STU
AVER

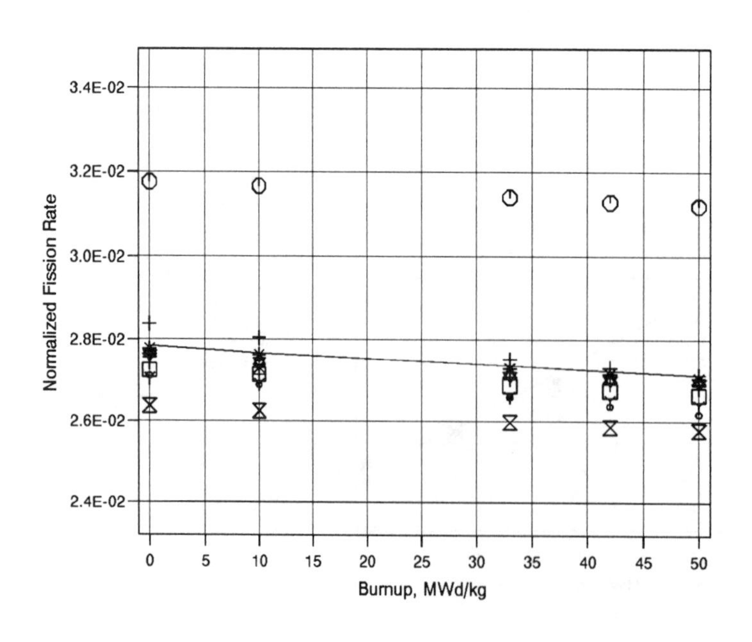
Benchmark A: Fission Rate
3.4E-02
3.2E-02
3.0E-02
2.8E-02
2.6E-02
2.4E-02
Normalized Fission Rate
0
5
10
15
20
25
30
35
40
45
50
Burnup, MWd/kg
U-238
BEN
BNFL
CEA
ECN
EDF
HIT
IKE1
JAE
PSI1
STU
AVER

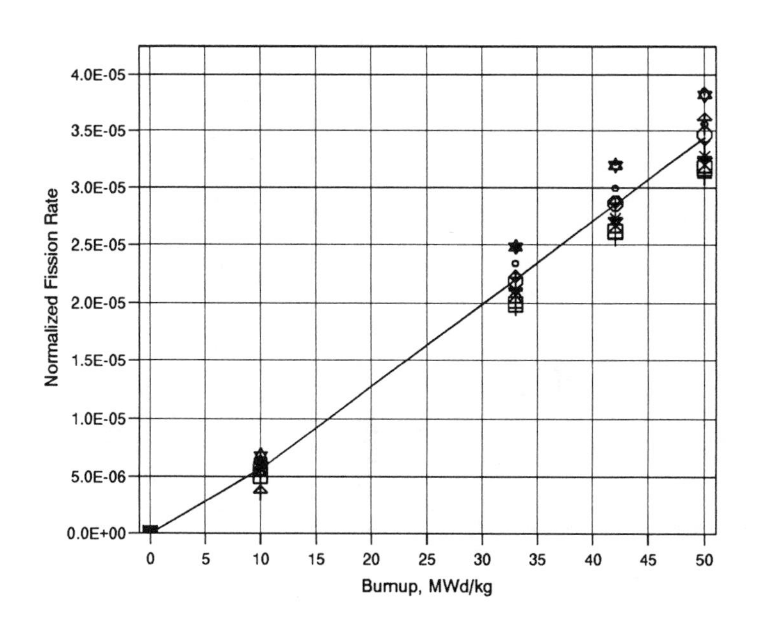
Benchmark A: Fission Rate
Normalized Fission Rate
4.0E-05
3.5E-05
3.0E-05
2.5E-05
2.0E-05
1.5E-05
1.0E-05
5.0E-06
0.0E+00
0
5
10
15
20
25
30
35
40
45
50
Burnup, MWd/kg
Np-237
BEN
BNFL
CEA
ECN
EDF
HIT
IKE1
JAE
PSI1
STU
AVER

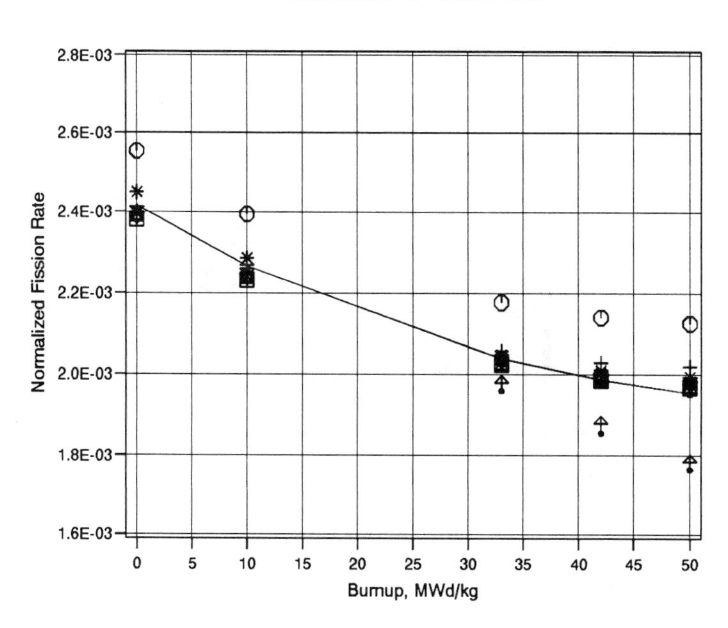
Benchmark A: Fission Rate
Normalized Fission Rate
2.8E-03
2.6E-03
2.4E-03
2.2E-03
2.0E-03
1.8E-03
1.6E-03
0
5
10
15
20
25
30
35
40
45
50
Burnup, MWd/kg
Pu-238
BEN
BNFL
CEA
ECN
EDF
HIT
IKE1
JAE
PSI1
STU
AVER

A-fr

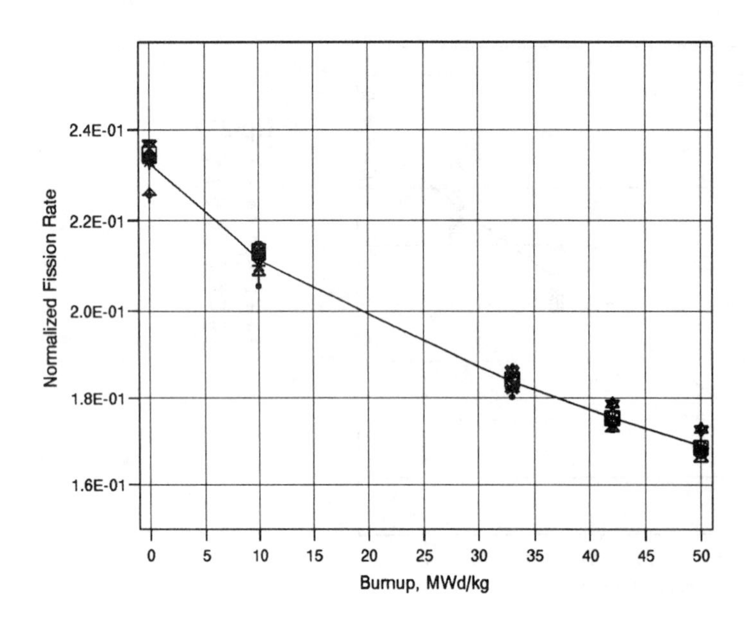

Benchmark A: Fission Rate
Normalized Fission Rate
2.4E-01
2.2E-01
2.0E-01
1.8E-01
1.6E-01
0
5
10
15
20
25
30
35
40
45
50
Burnup, MWd/kg
Pu-239
BEN
BNFL
CEA
ECN
EDF
HIT
IKE1
JAE
PSI1
STU
AVER

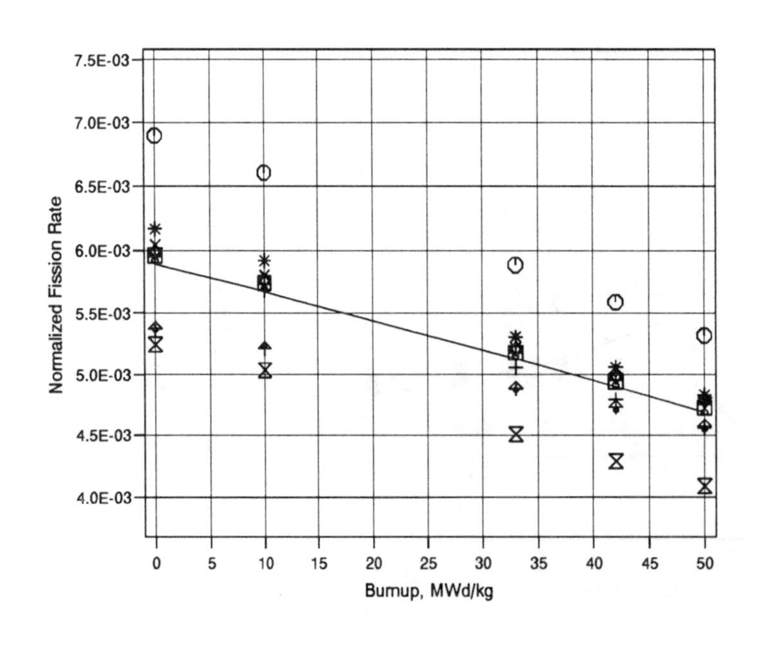

Benchmark A: Fission Rate
Normalized Fission Rate
7.5E-03
7.0E-03
6.5E-03
6.0E-03
5.5E-03
5.0E-03
4.5E-03
4.0E-03
0
5
10
15
20
25
30
35
40
45
50
Burnup, MWd/kg
Pu-240
BEN
BNFL
CEA
ECN
EDF
HIT
IKE1
JAE
PSI1
STU
AVER

Benchmark A: Fission Rate

A-fr

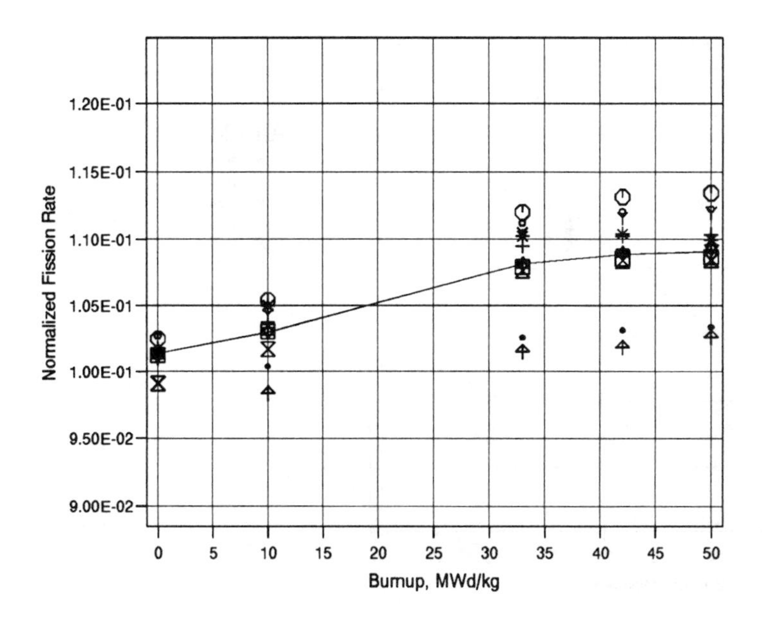

Pu-241

- BEN
- BNFL
- CEA
- ECN
- EDF
- HIT
- IKE1
- JAE
- PSI1
- STU
- AVER

Benchmark A: Fission Rate

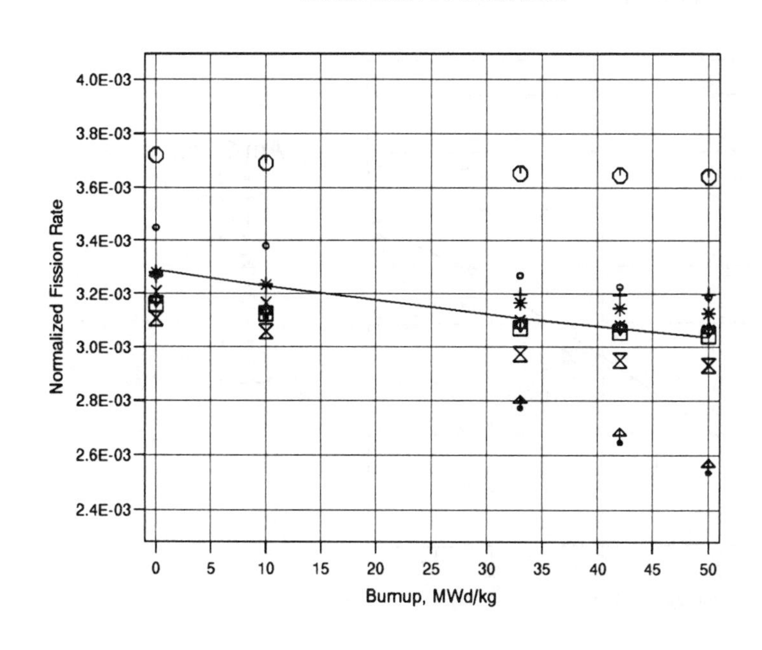

Pu-242

- BEN
- BNFL
- CEA
- ECN
- EDF
- HIT
- IKE1
- JAE
- PSI1
- STU
- AVER

A-fr

Benchmark A: Fission Rate

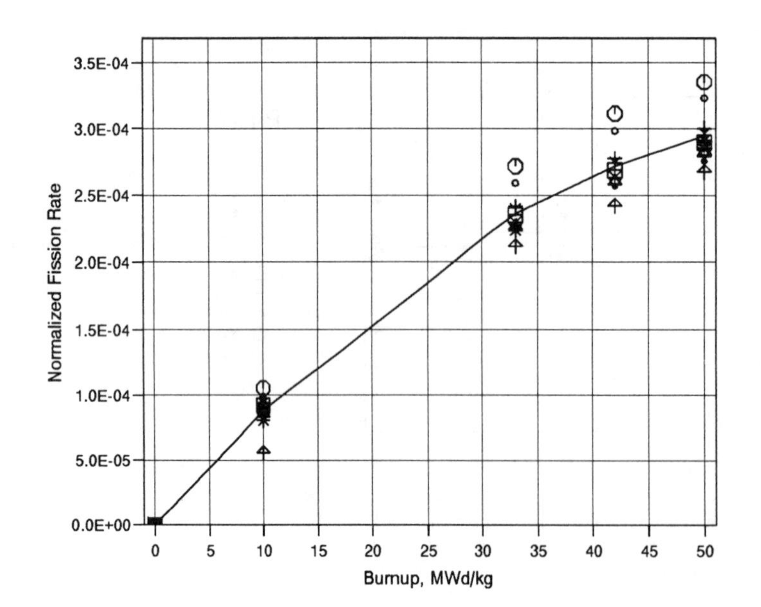

Am-241

- BEN
- BNFL
- CEA
- ECN
- EDF
- HIT
- IKE1
- JAE
- PSI1
- STU
- AVER

Benchmark A: Fission Rate

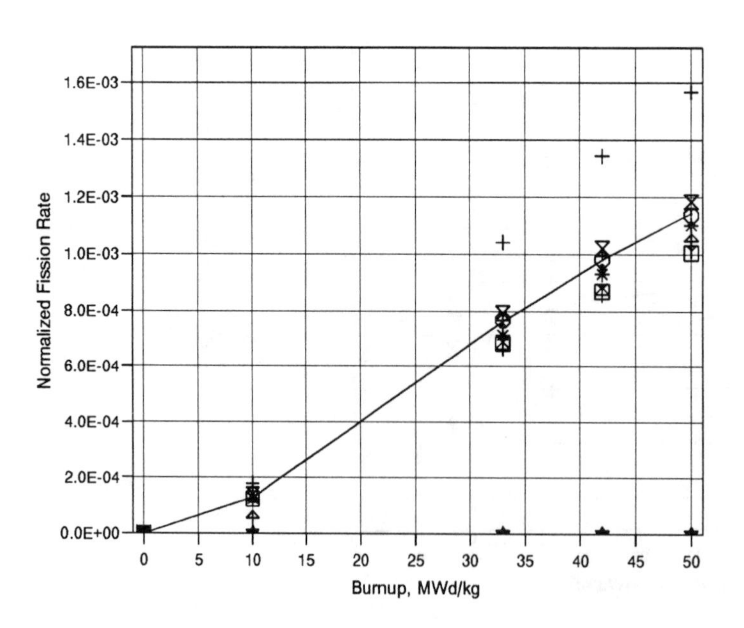

Am-242m

- BEN
- BNFL
- CEA
- ECN
- EDF
- HIT
- IKE1
- JAE
- PSI1
- STU
- AVER

Benchmark A: Fission Rate

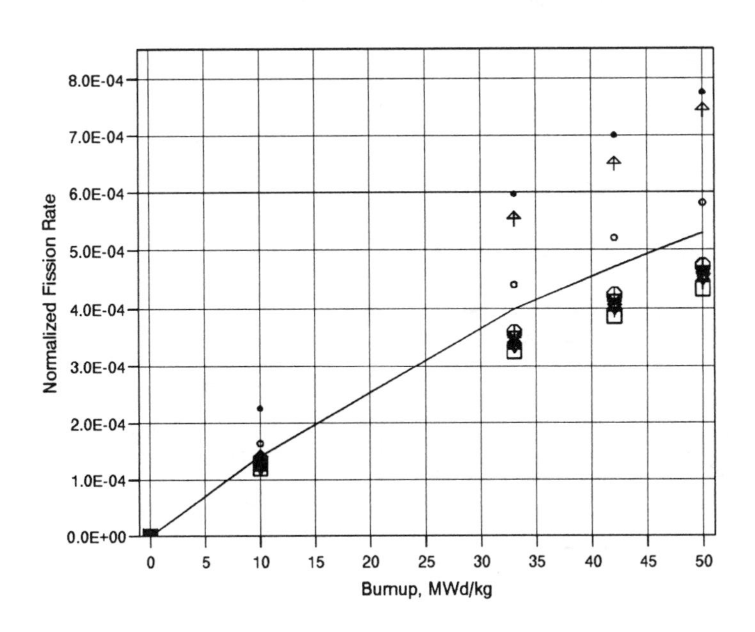

Am-243

- BEN
- BNFL
- CEA
- ECN
- EDF
- HIT
- IKE1
- JAE
- PSI1
- STU
- AVER

Benchmark A: Fission Rate

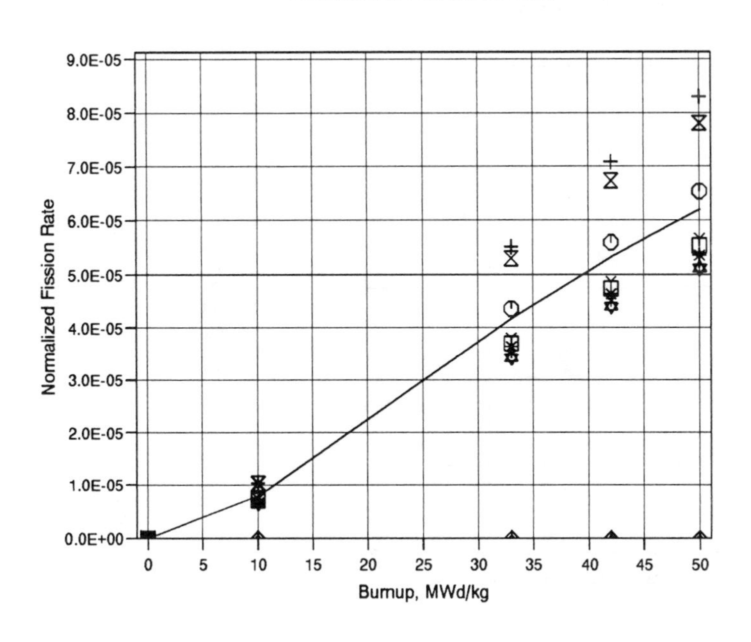

Cm-242

A-fr

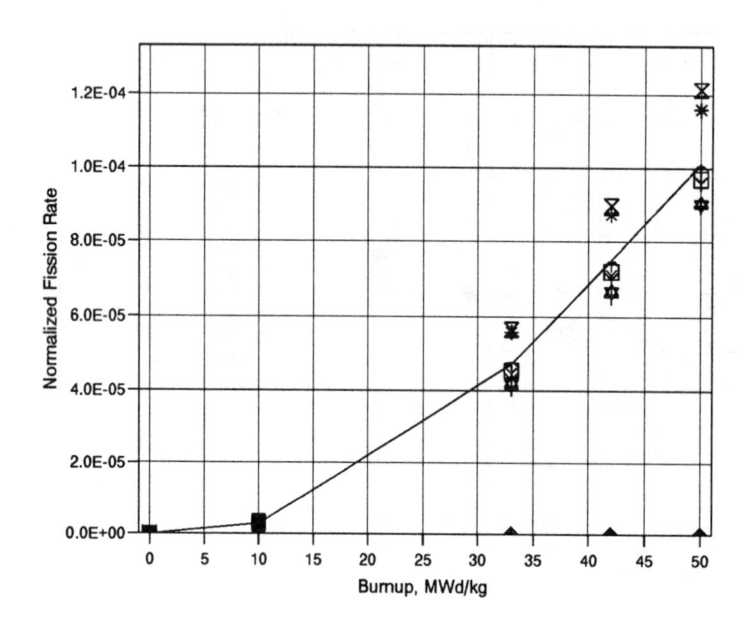
Benchmark A: Fission Rate
Normalized Fission Rate
1.2E-04
1.0E-04
8.0E-05
6.0E-05
4.0E-05
2.0E-05
0.0E+00
0
5
10
15
20
25
30
35
40
45
50
Burnup, MWd/kg
Cm-243
BEN
BNFL
CEA
ECN
EDF
HIT
IKE1
JAE
PSI1
STU
AVER

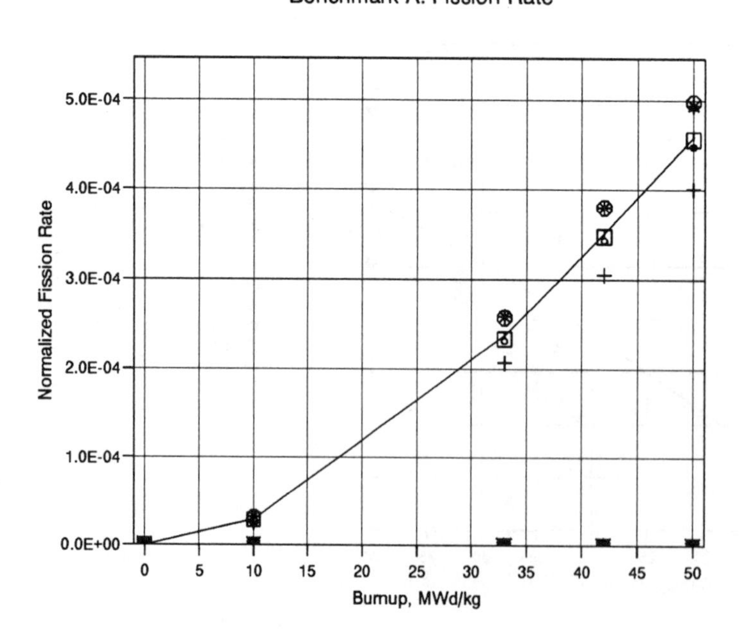
Benchmark A: Fission Rate
Normalized Fission Rate
5.0E-04
4.0E-04
3.0E-04
2.0E-04
1.0E-04
0.0E+00
0
5
10
15
20
25
30
35
40
45
50
Burnup, MWd/kg
Cm-244
BEN
BNFL
CEA
ECN
EDF
HIT
IKE1
JAE
PSI1
STU
AVER

Benchmark A: Fission Rate

A-fr

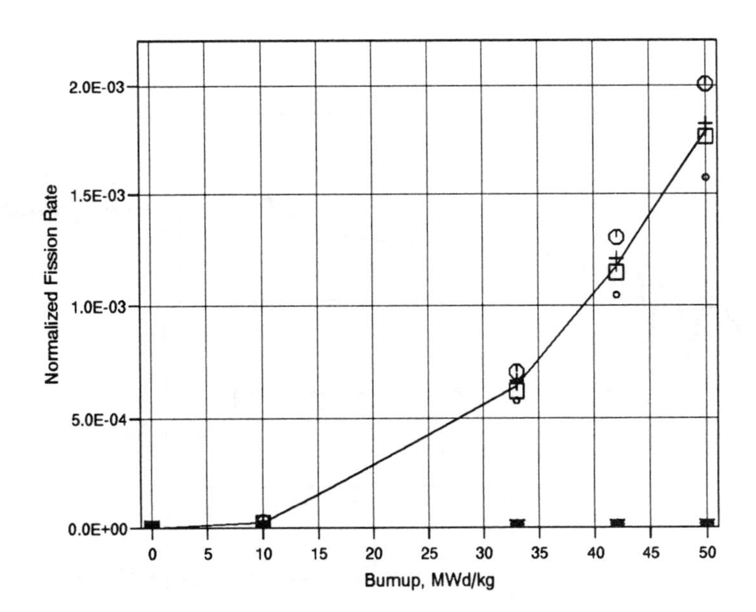

A Contributor: BNFL

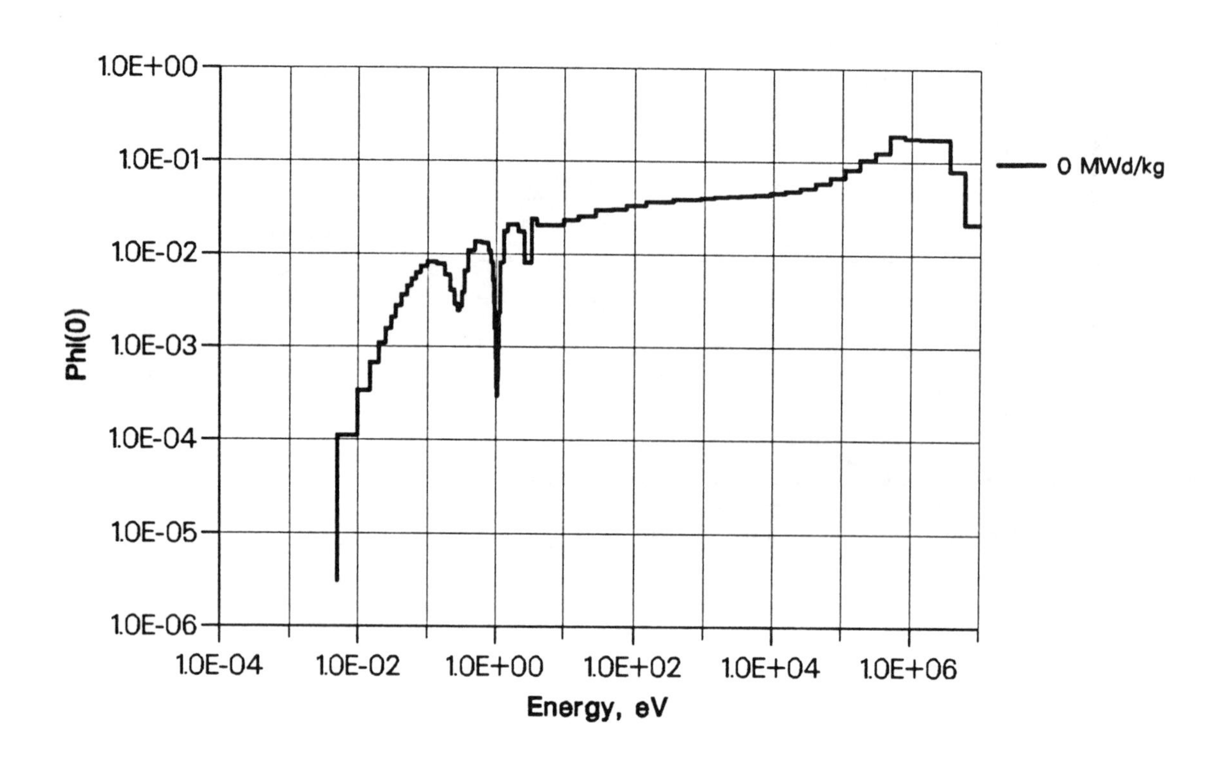

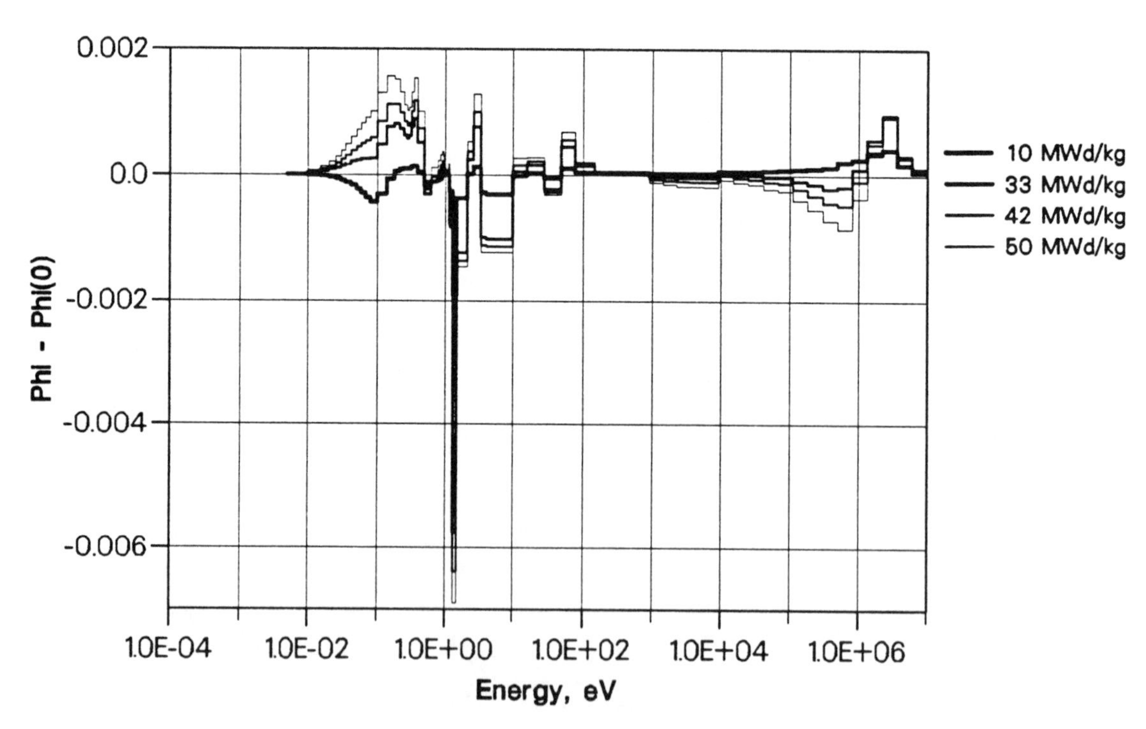

Benchmark 1-A SPECTRA as a function of burnup

Contributor: CEA

A

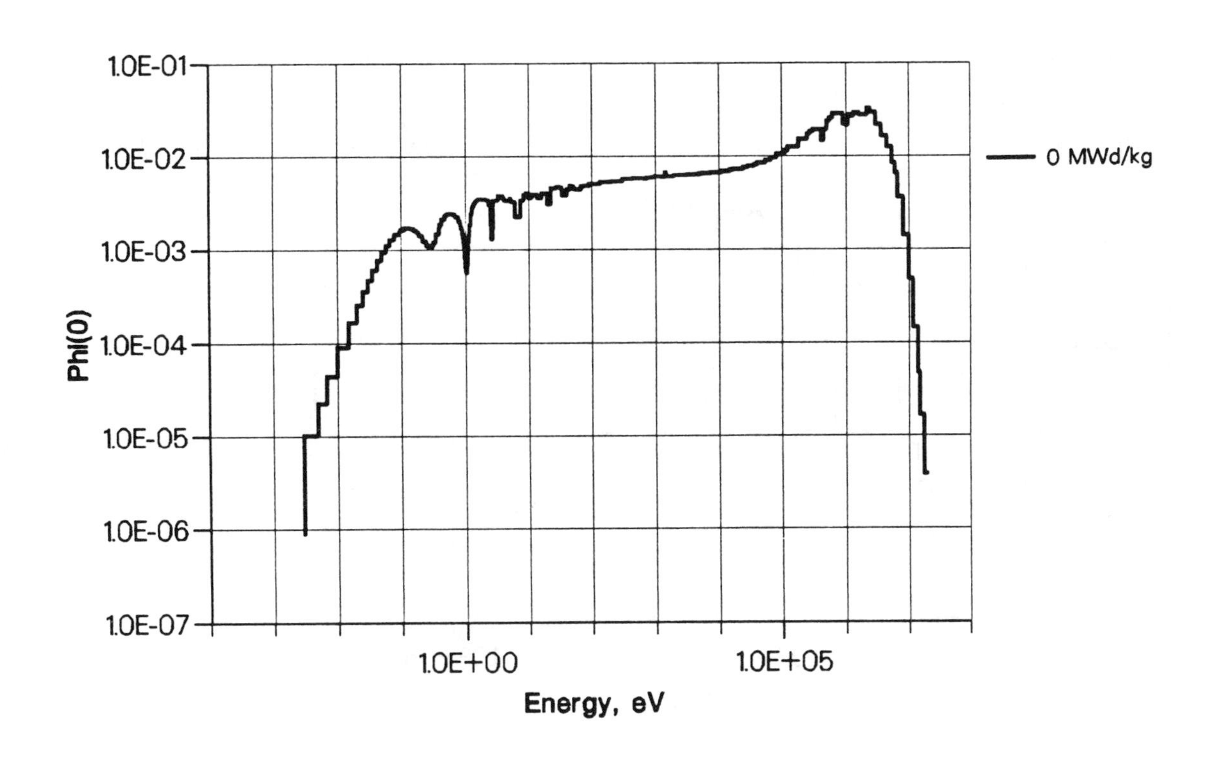

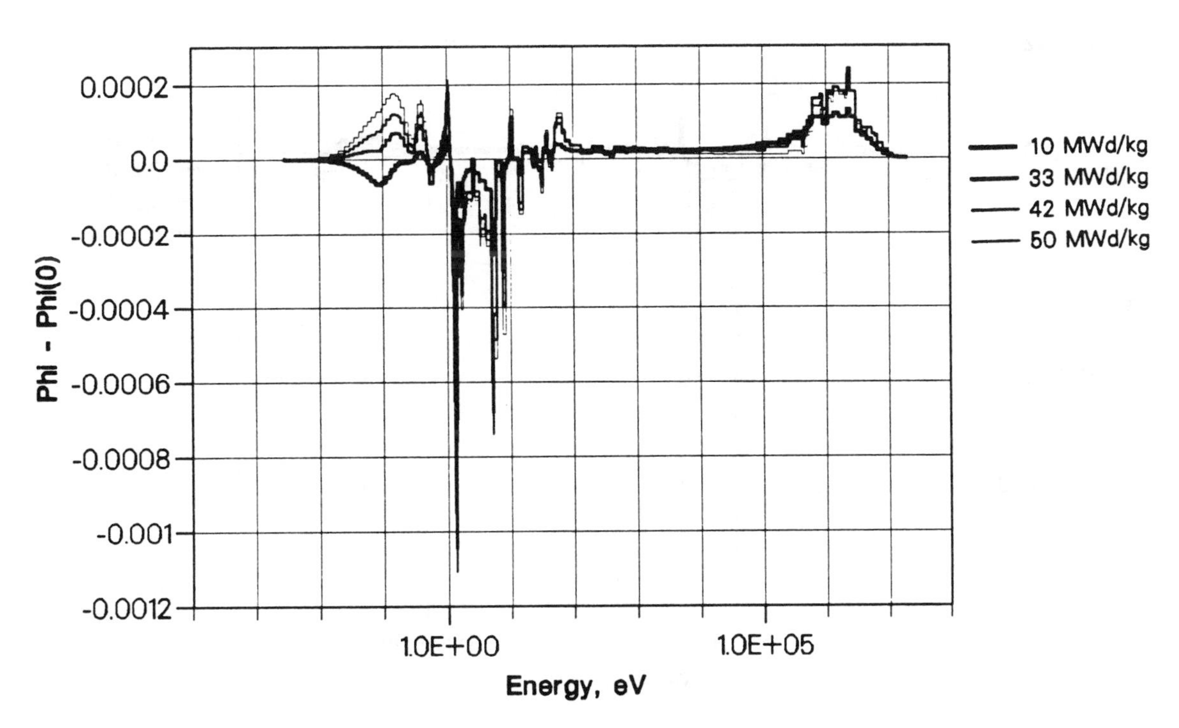

Benchmark 1-A SPECTRA as a function of burnup

A

Contributor: ECN

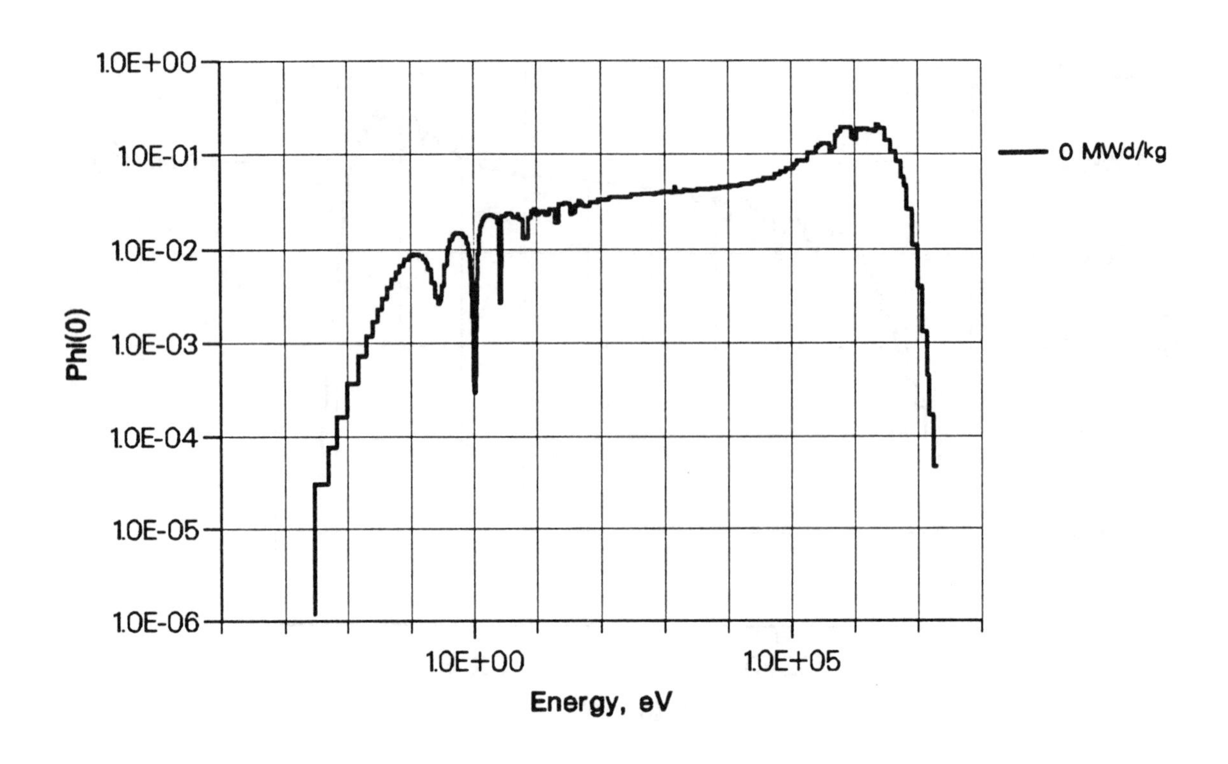

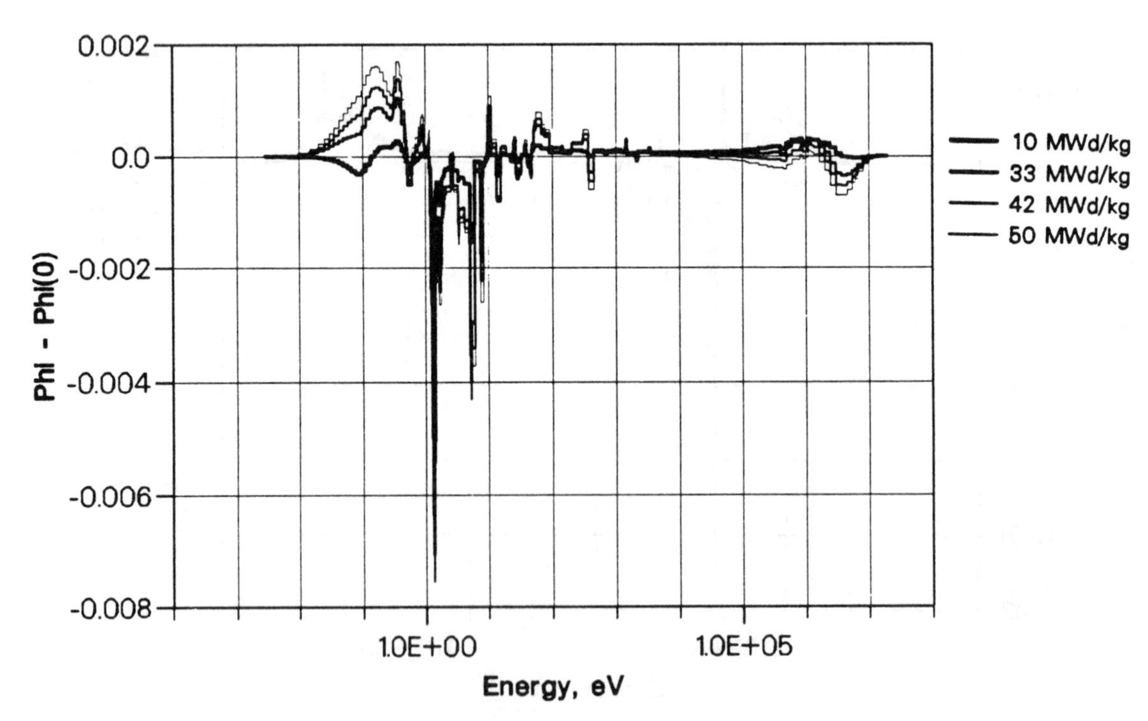

Benchmark 1-A SPECTRA as a function of burnup

Contributor: EDF

A

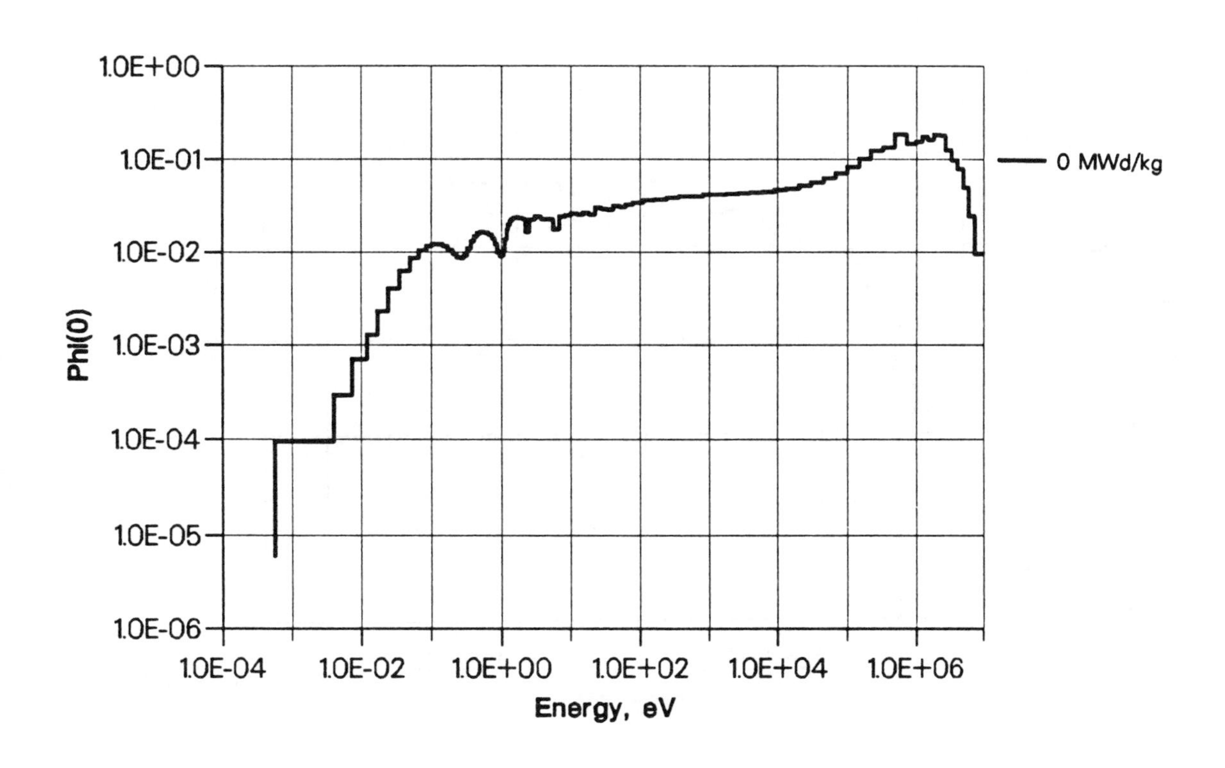

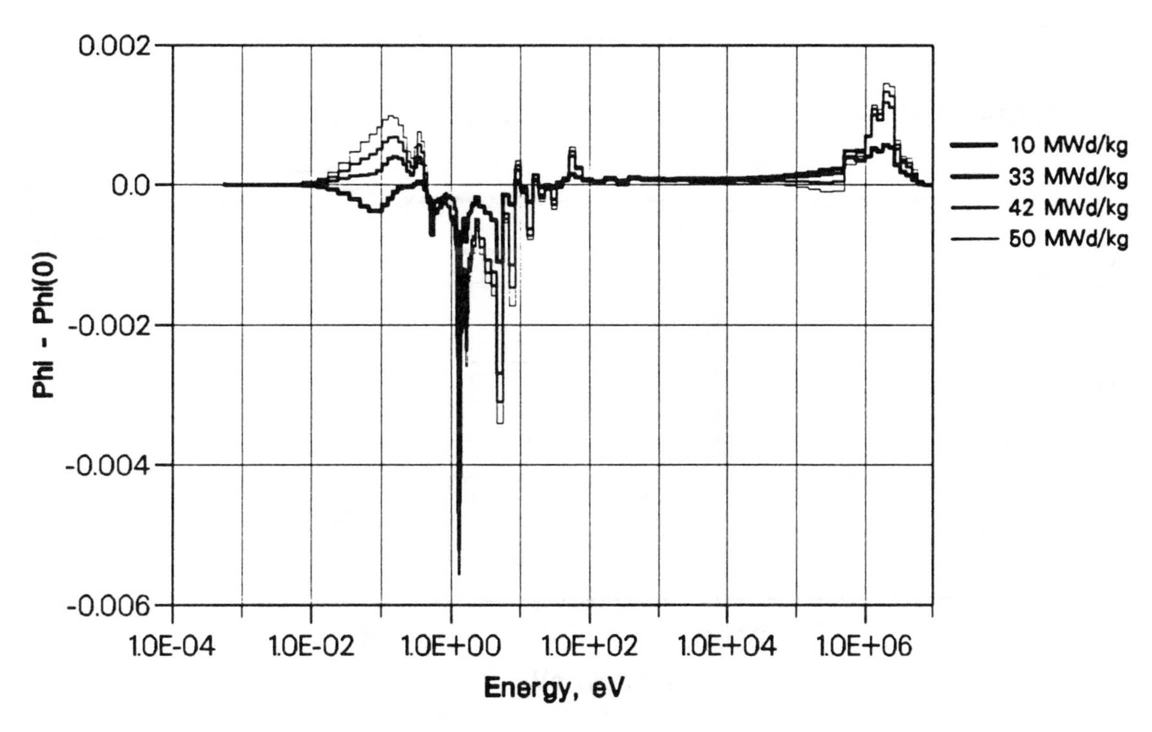

Benchmark 1-A SPECTRA as a function of burnup

A

Contributor: IKE 1

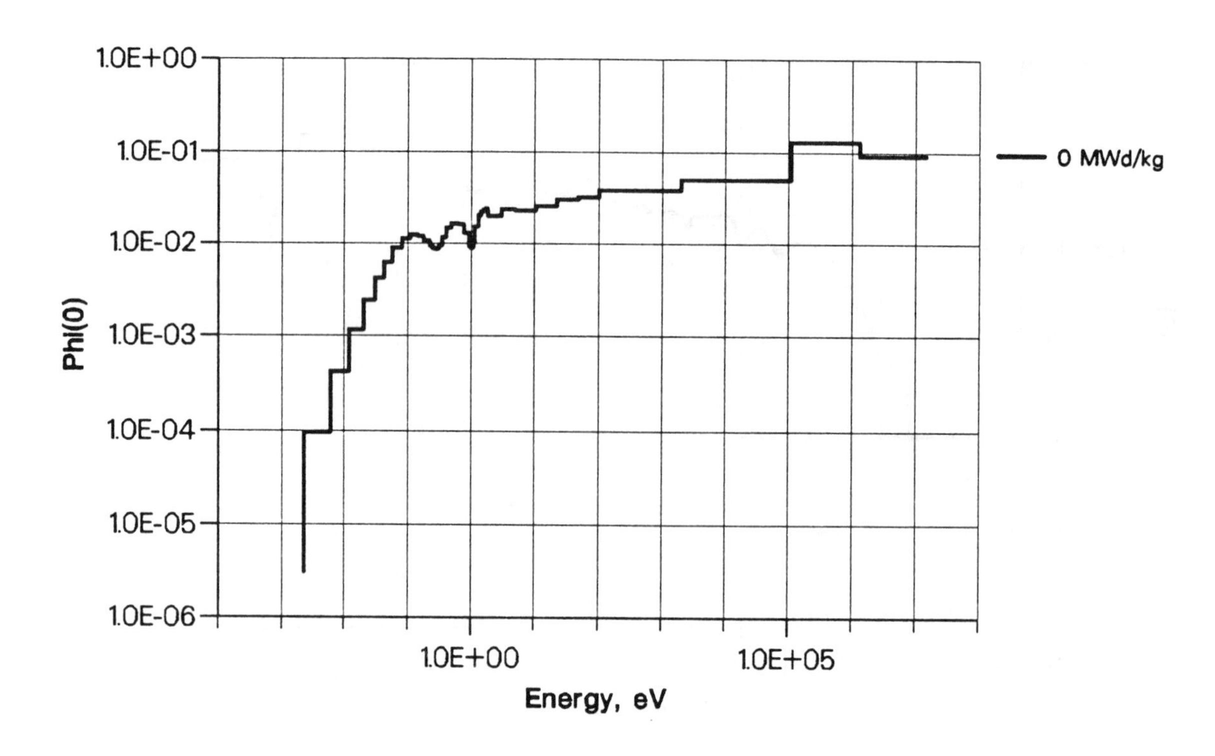

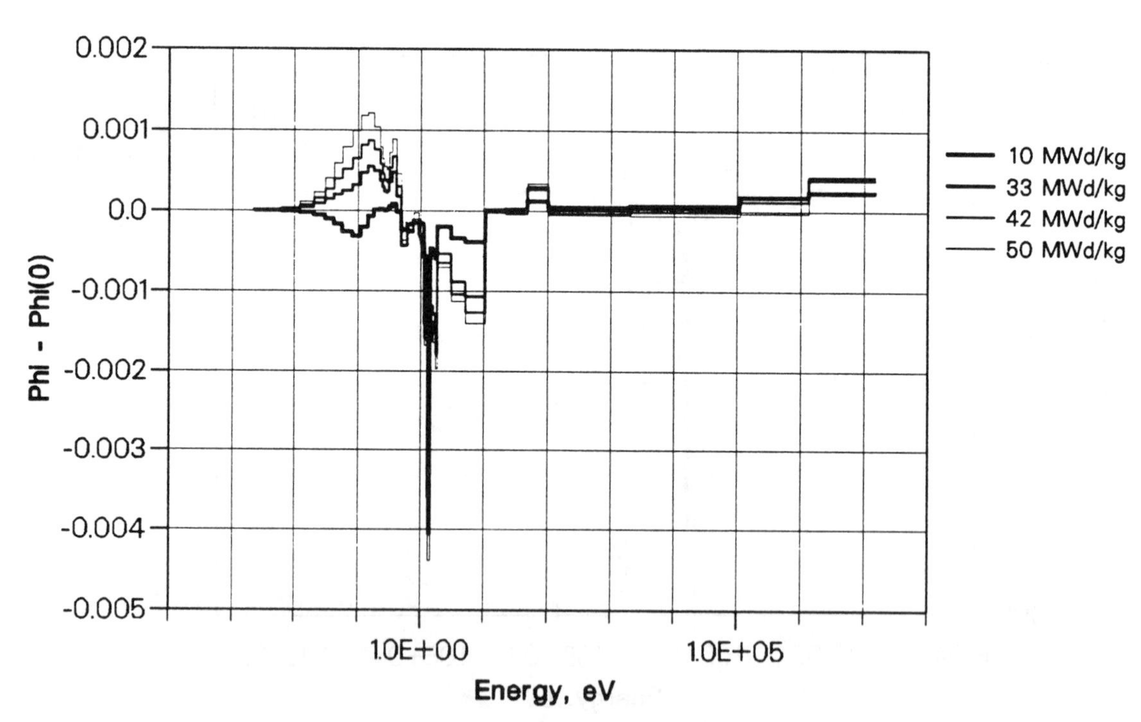

Benchmark 1-A SPECTRA as a function of burnup

Contributor: JAERI

A

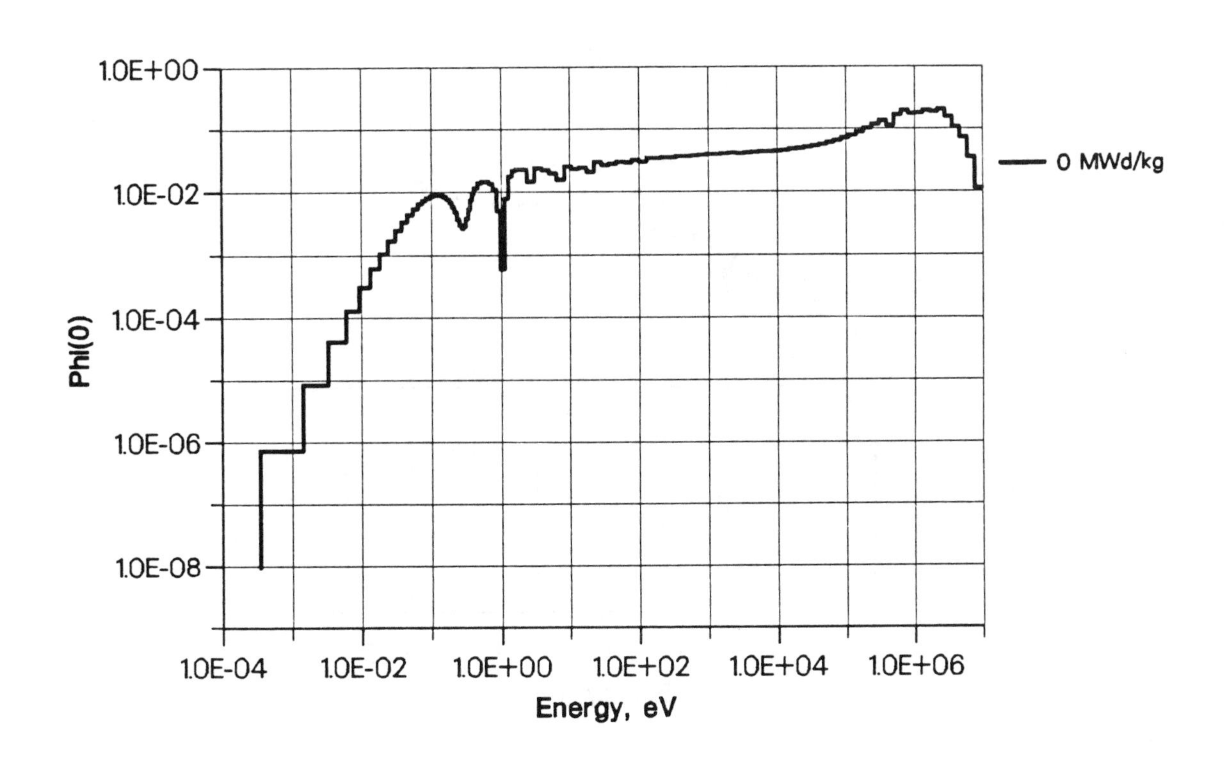

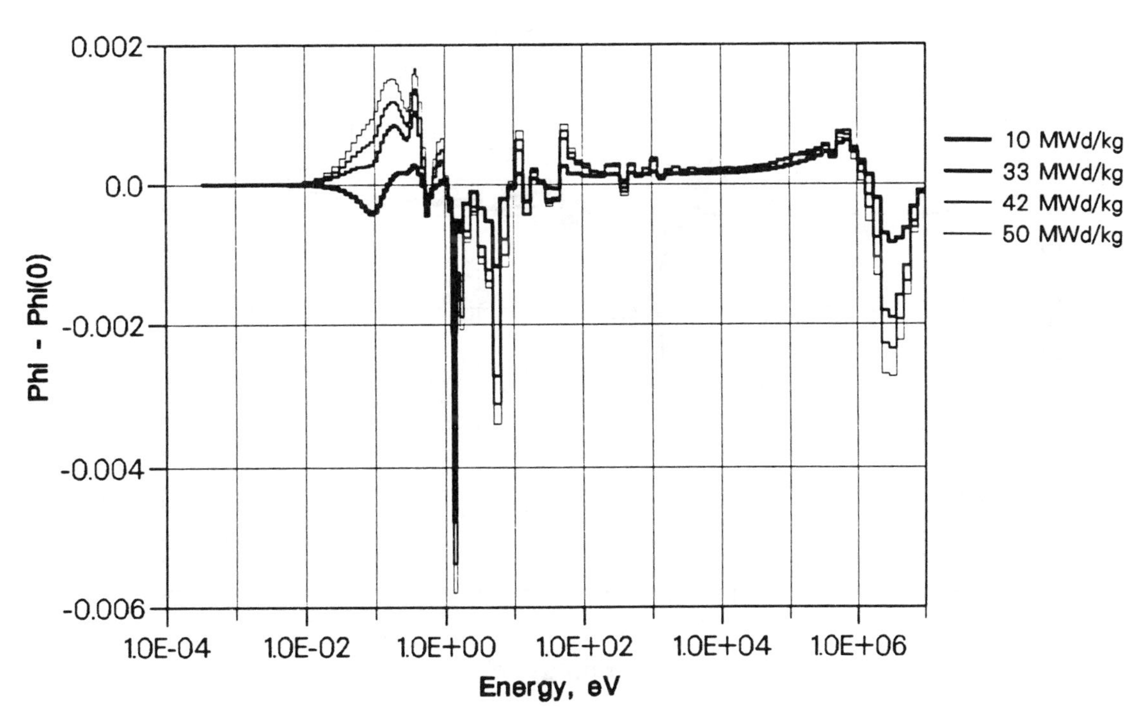

Benchmark 1-A SPECTRA as a function of burnup

A

Contributor: PSI 1

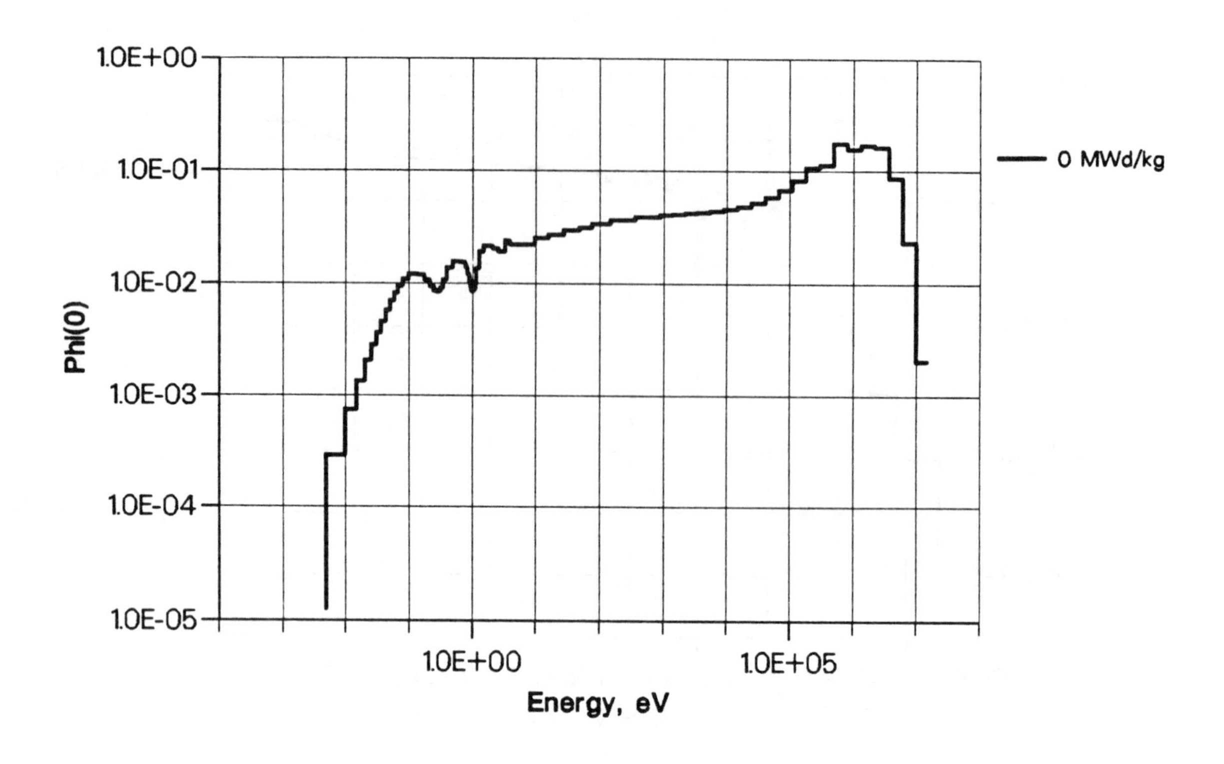

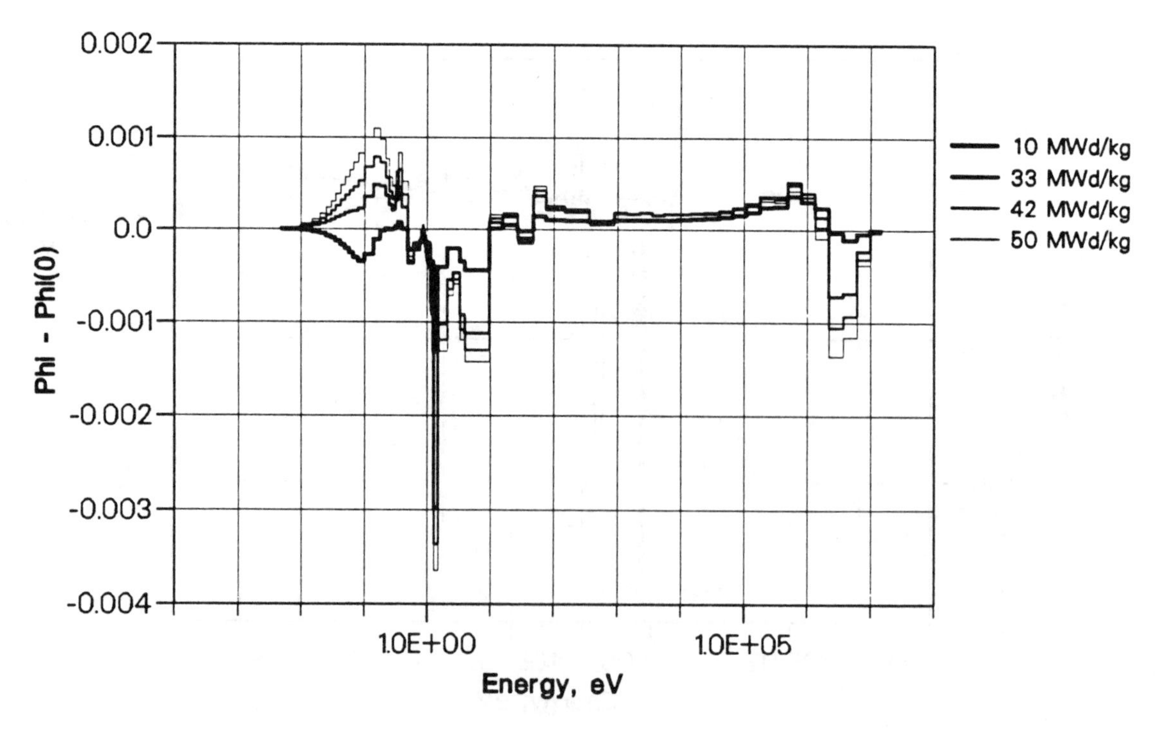

Benchmark 1-A SPECTRA as a function of burnup

B-ar

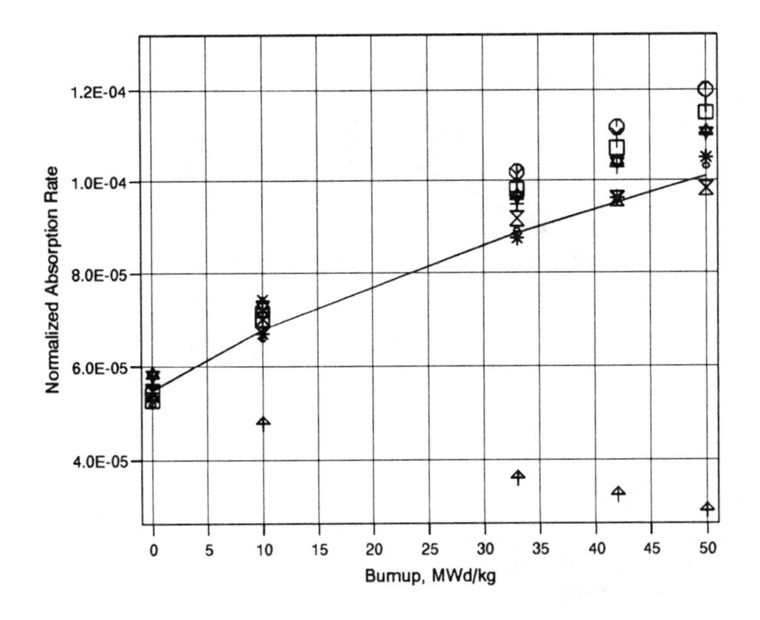

Benchmark B: Absorption Rate (Total normalized to 1)
1.2E-04
1.0E-04
8.0E-05
6.0E-05
4.0E-05
Normalized Absorption Rate
0
5
10
15
20
25
30
35
40
45
50
Burnup, MWd/kg
U-234
BNFL
CEA
ECN
EDF
HIT
IKE1
JAE
PSI1
STU
AVER

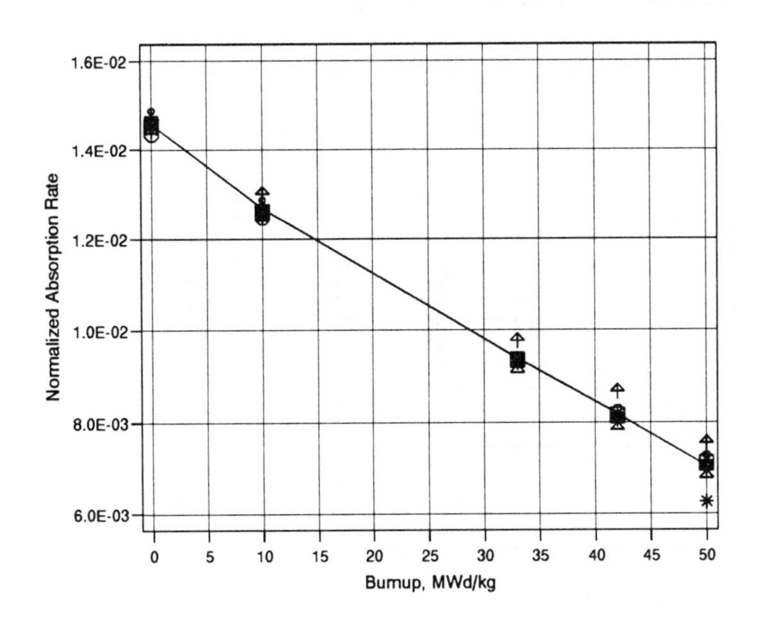

Benchmark B: Absorption Rate (Total normalized to 1)
1.6E-02
1.4E-02
1.2E-02
1.0E-02
8.0E-03
6.0E-03
Normalized Absorption Rate
0
5
10
15
20
25
30
35
40
45
50
Burnup, MWd/kg
U-235
BNFL
CEA
ECN
EDF
HIT
IKE1
JAE
PSI1
STU
AVER

B-ar

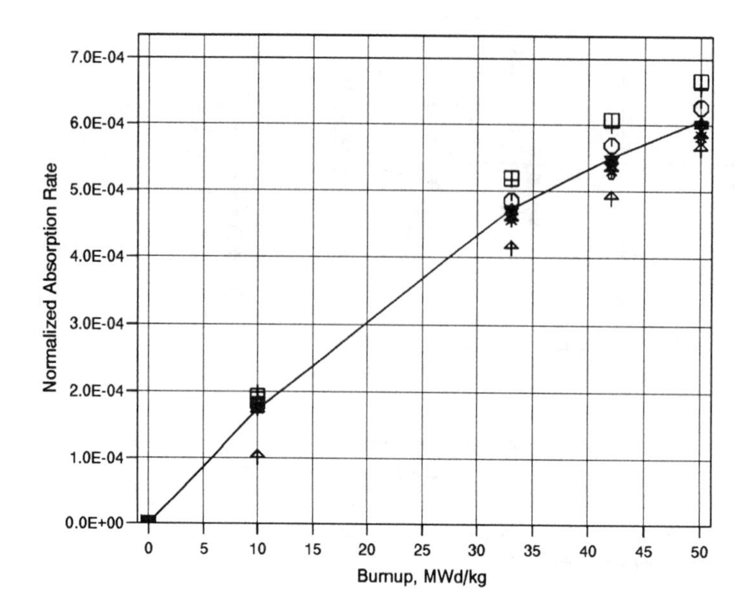
Benchmark B: Absorption Rate (Total normalized to 1)
Normalized Absorption Rate
7.0E-04
6.0E-04
5.0E-04
4.0E-04
3.0E-04
2.0E-04
1.0E-04
0.0E+00
0
5
10
15
20
25
30
35
40
45
50
Burnup, MWd/kg
U-236
BNFL
CEA
ECN
EDF
HIT
IKE1
JAE
PSI1
STU
AVER

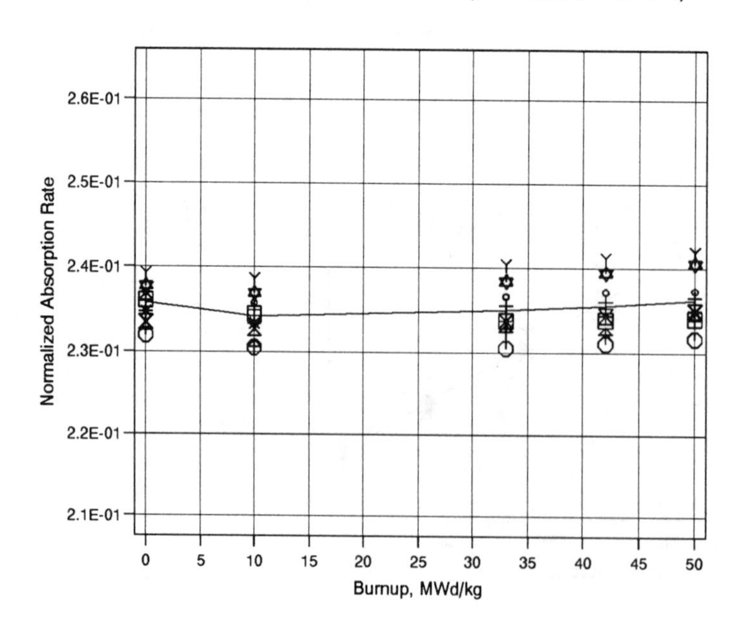
Benchmark B: Absorption Rate (Total normalized to 1)
Normalized Absorption Rate
2.6E-01
2.5E-01
2.4E-01
2.3E-01
2.2E-01
2.1E-01
0
5
10
15
20
25
30
35
40
45
50
Burnup, MWd/kg
U-238
BNFL
CEA
ECN
EDF
HIT
IKE1
JAE
PSI1
STU
AVER

Benchmark B: Absorption Rate (Total normalized to 1)

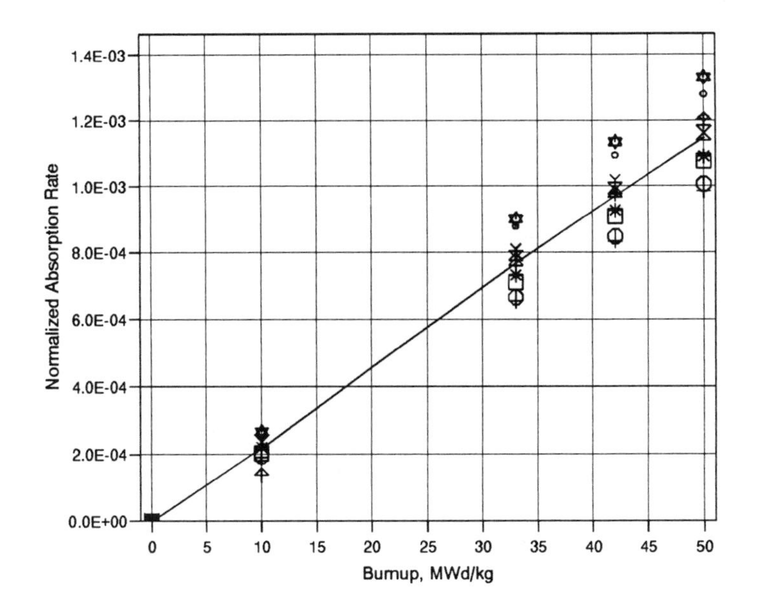

Benchmark B: Absorption Rate (Total normalized to 1)

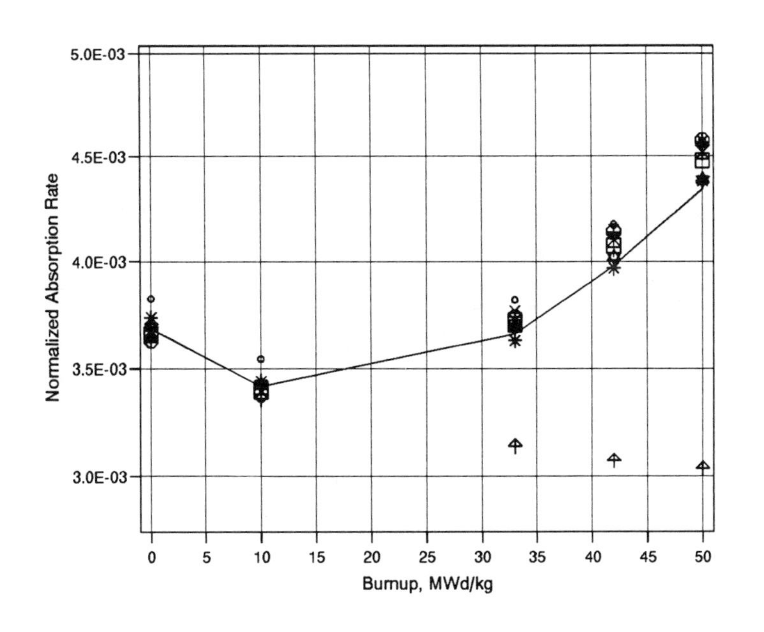

B-ar

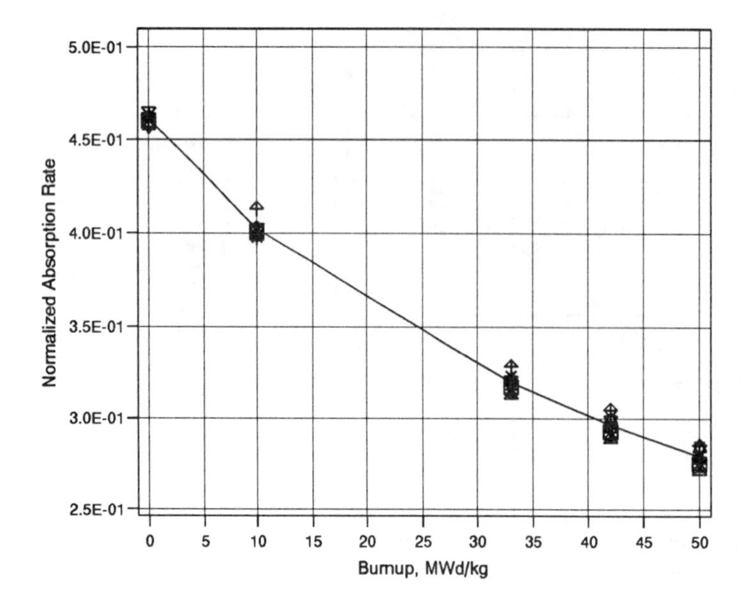
Benchmark B: Absorption Rate (Total normalized to 1)
Normalized Absorption Rate
5.0E-01
4.5E-01
4.0E-01
3.5E-01
3.0E-01
2.5E-01
0
5
10
15
20
25
30
35
40
45
50
Burnup, MWd/kg
Pu-239
BNFL
CEA
ECN
EDF
HIT
IKE1
JAE
PSI1
STU
AVER

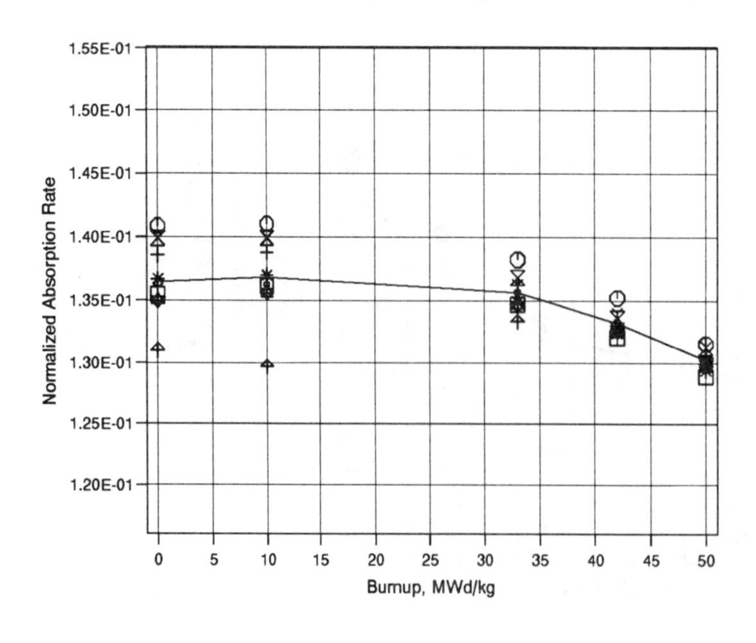
Benchmark B: Absorption Rate (Total normalized to 1)
Normalized Absorption Rate
1.55E-01
1.50E-01
1.45E-01
1.40E-01
1.35E-01
1.30E-01
1.25E-01
1.20E-01
0
5
10
15
20
25
30
35
40
45
50
Burnup, MWd/kg
Pu-240
BNFL
CEA
ECN
EDF
HIT
IKE1
JAE
PSI1
STU
AVER

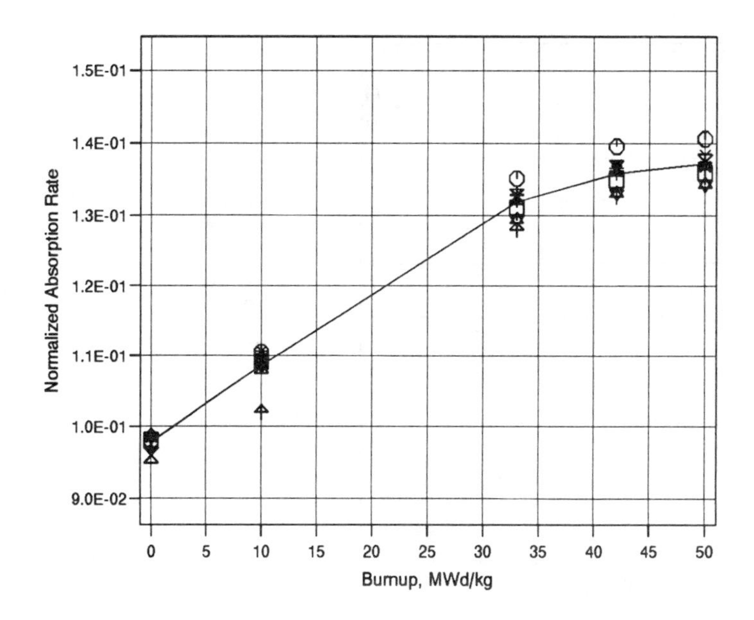
Benchmark B: Absorption Rate (Total normalized to 1)
Normalized Absorption Rate
1.5E-01
1.4E-01
1.3E-01
1.2E-01
1.1E-01
1.0E-01
9.0E-02
0
5
10
15
20
25
30
35
40
45
50
Burnup, MWd/kg
Pu-241
BNFL
CEA
ECN
EDF
HIT
IKE1
JAE
PSI1
STU
AVER

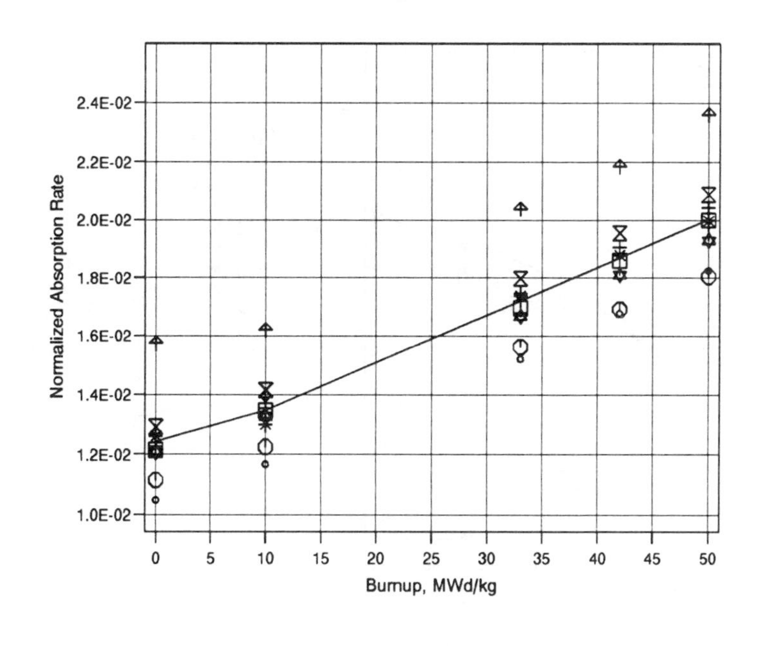
Benchmark B: Absorption Rate (Total normalized to 1)
Normalized Absorption Rate
2.4E-02
2.2E-02
2.0E-02
1.8E-02
1.6E-02
1.4E-02
1.2E-02
1.0E-02
0
5
10
15
20
25
30
35
40
45
50
Burnup, MWd/kg
Pu-242
BNFL
CEA
ECN
EDF
HIT
IKE1
JAE
PSI1
STU
AVER

B-ar

Benchmark B: Absorption Rate (Total normalized to 1)

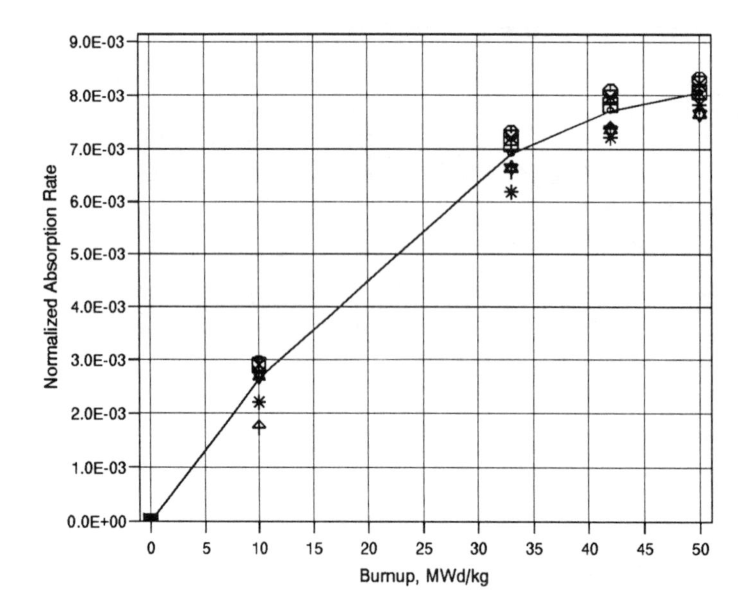

Am-241
BNFL
CEA
ECN
EDF
HIT
IKE1
JAE
PSI1
STU
AVER

Benchmark B: Absorption Rate (Total normalized to 1)

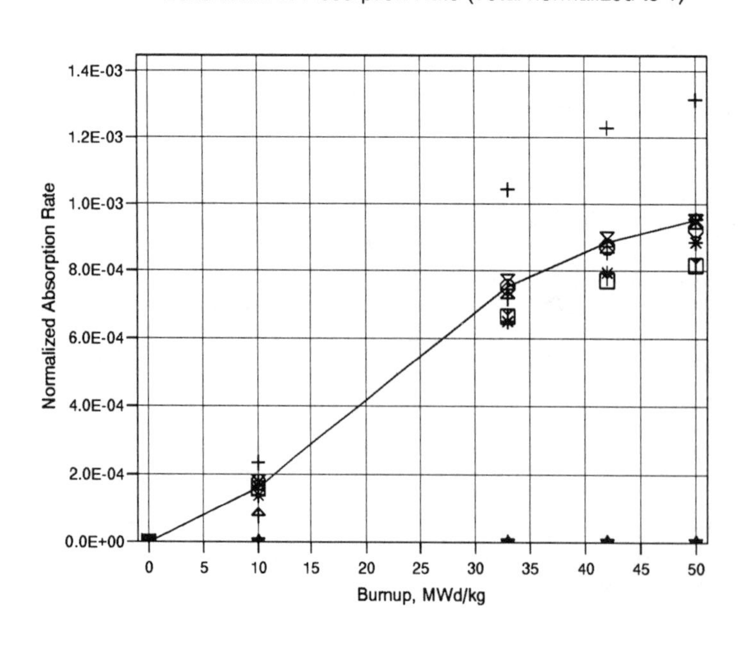

Am-242m

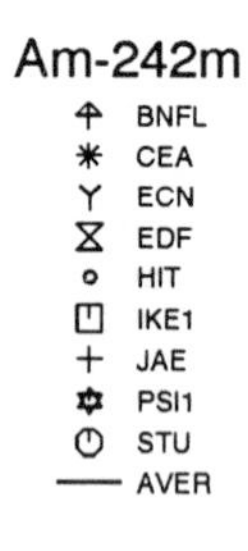

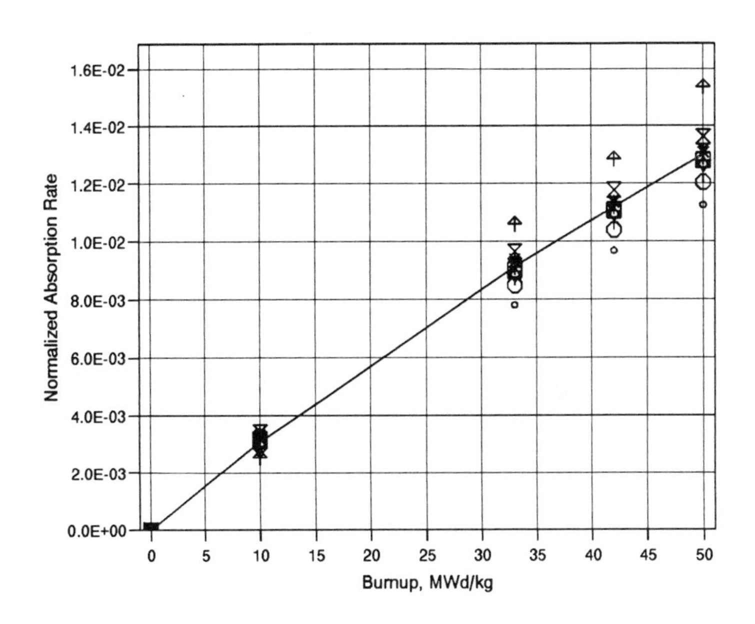
Benchmark B: Absorption Rate (Total normalized to 1)
Normalized Absorption Rate
Burnup, MWd/kg
1.6E-02
1.4E-02
1.2E-02
1.0E-02
8.0E-03
6.0E-03
4.0E-03
2.0E-03
0.0E+00
0
5
10
15
20
25
30
35
40
45
50
Am-243
BNFL
CEA
ECN
EDF
HIT
IKE1
JAE
PSI1
STU
AVER

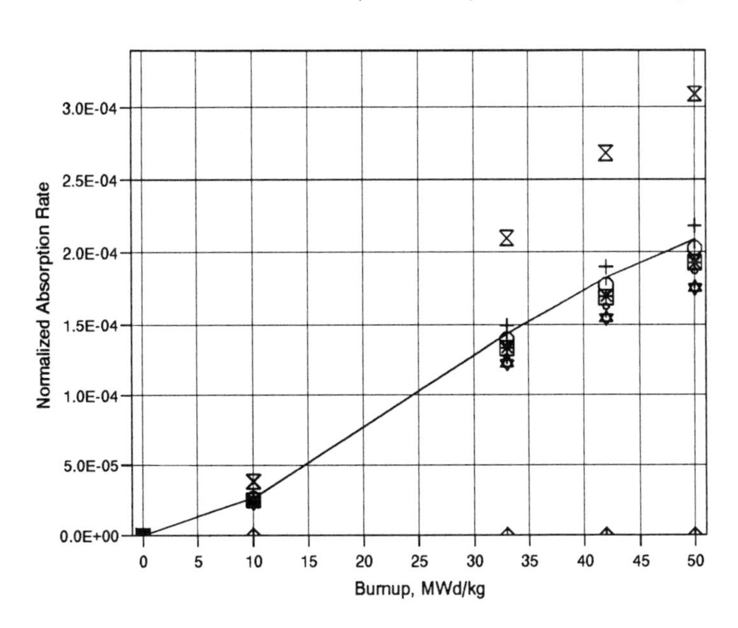
Benchmark B: Absorption Rate (Total normalized to 1)
Normalized Absorption Rate
Burnup, MWd/kg
3.0E-04
2.5E-04
2.0E-04
1.5E-04
1.0E-04
5.0E-05
0.0E+00
0
5
10
15
20
25
30
35
40
45
50
Cm-242
BNFL
CEA
ECN
EDF
HIT
IKE1
JAE
PSI1
STU
AVER

Benchmark B: Absorption Rate (Total normalized to 1)

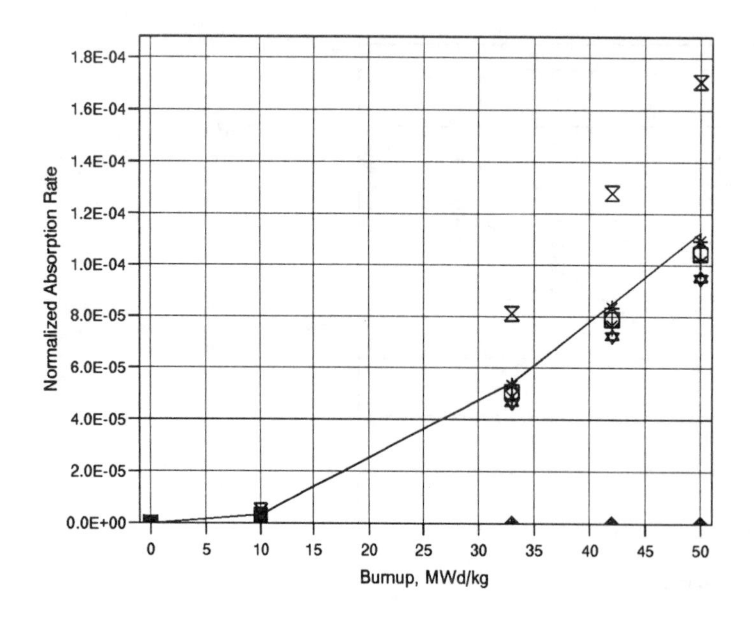

Benchmark B: Absorption Rate (Total normalized to 1)

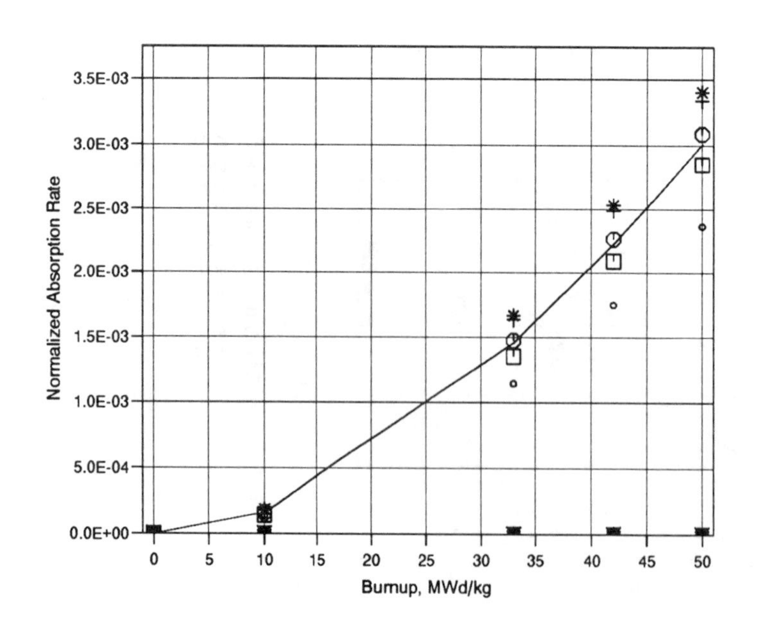

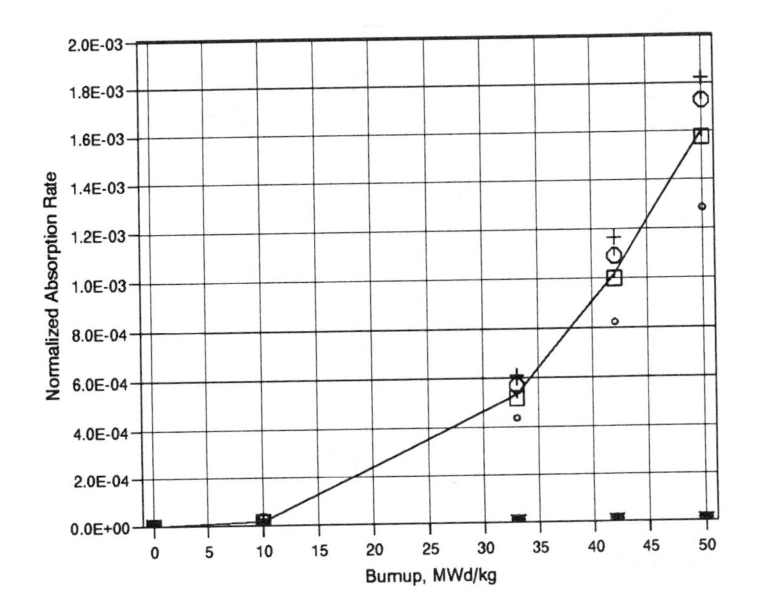
Benchmark B: Absorption Rate (Total normalized to 1)
Normalized Absorption Rate
2.0E-03
1.8E-03
1.6E-03
1.4E-03
1.2E-03
1.0E-03
8.0E-04
6.0E-04
4.0E-04
2.0E-04
0.0E+00
0
5
10
15
20
25
30
35
40
45
50
Burnup, MWd/kg
Cm-245
BNFL
CEA
ECN
EDF
HIT
IKE1
JAE
PSI1
STU
AVER

B-fr

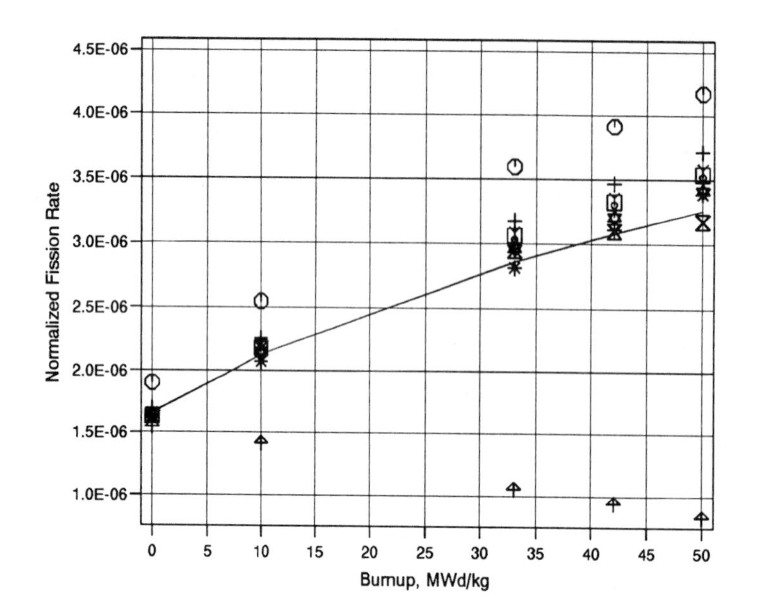

Benchmark B: Fission Rate
Normalized Fission Rate
4.5E-06
4.0E-06
3.5E-06
3.0E-06
2.5E-06
2.0E-06
1.5E-06
1.0E-06
0
5
10
15
20
25
30
35
40
45
50
Burnup, MWd/kg
U-234
BNFL
CEA
ECN
EDF
HIT
IKE1
JAE
PSI1
STU
AVER

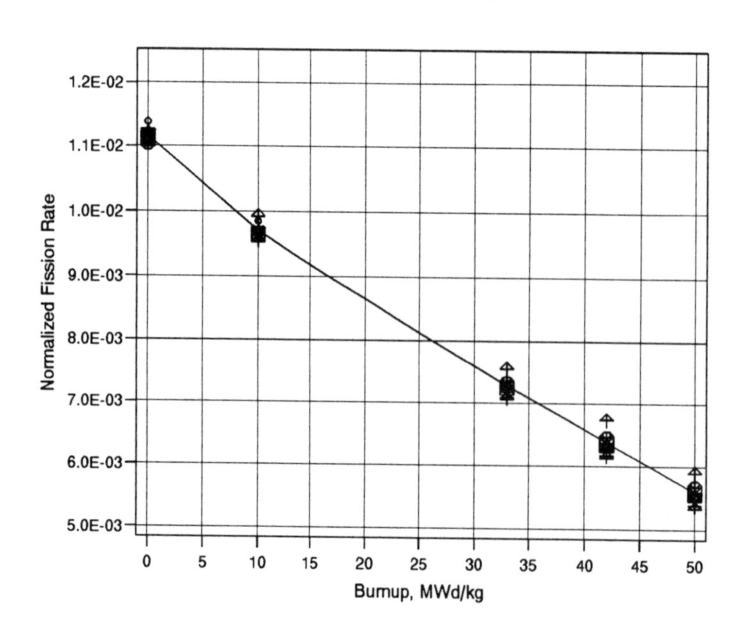

Benchmark B: Fission Rate
Normalized Fission Rate
1.2E-02
1.1E-02
1.0E-02
9.0E-03
8.0E-03
7.0E-03
6.0E-03
5.0E-03
0
5
10
15
20
25
30
35
40
45
50
Burnup, MWd/kg

U-235
BNFL
CEA
ECN
EDF
HIT
IKE1
JAE
PSI1
STU
AVER

Benchmark B: Fission Rate

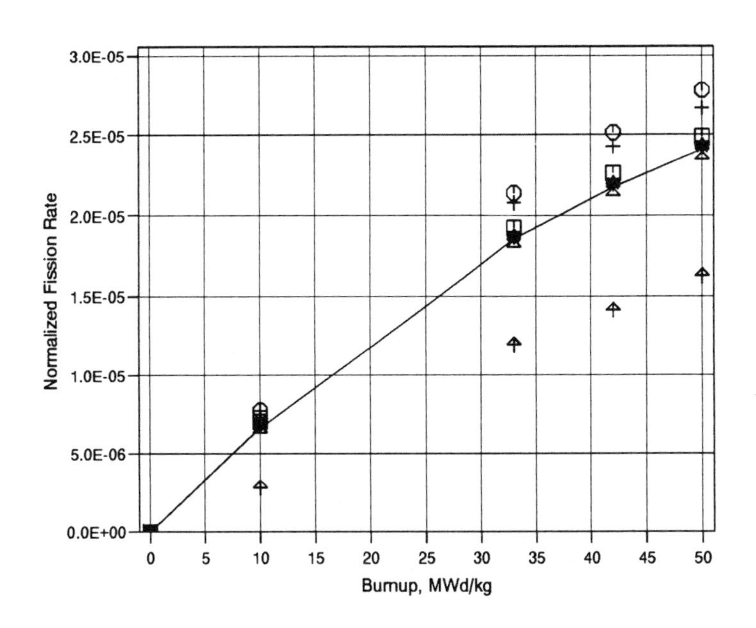

Benchmark B: Fission Rate

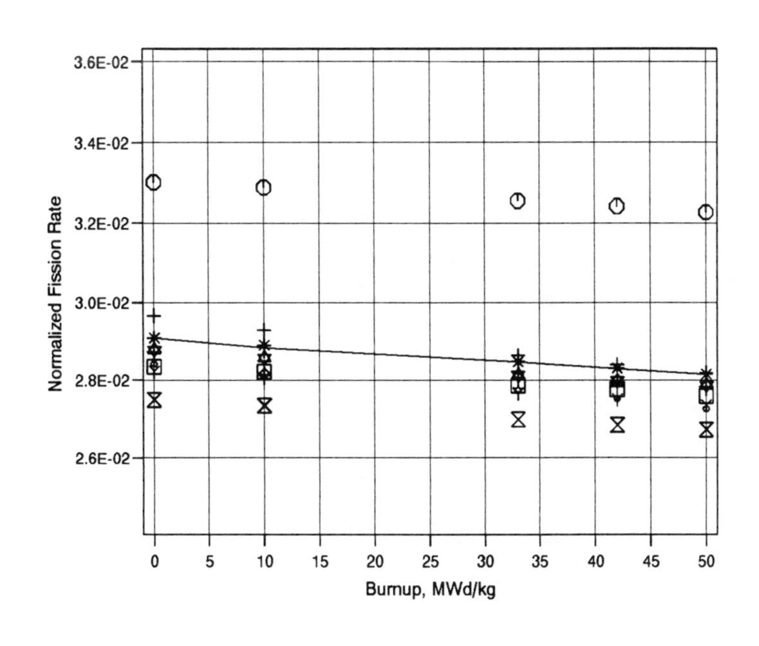

B-fr

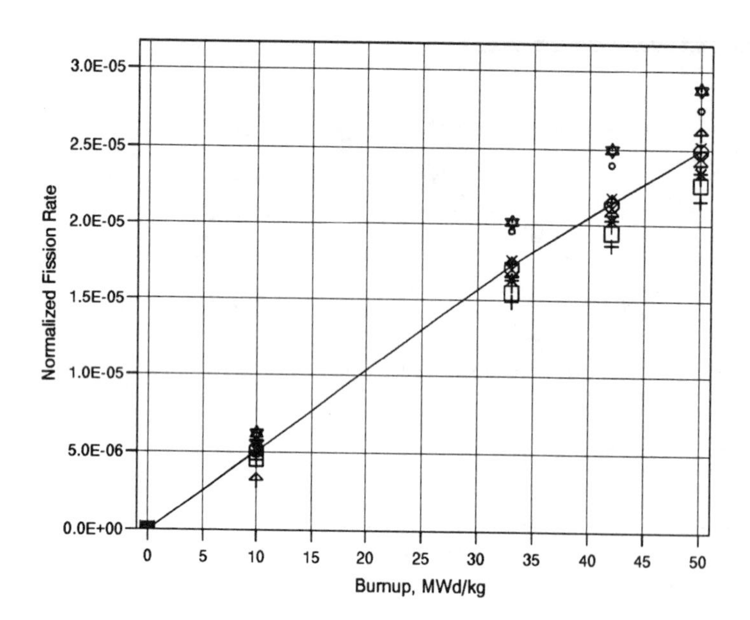

Benchmark B: Fission Rate
Normalized Fission Rate
3.0E-05
2.5E-05
2.0E-05
1.5E-05
1.0E-05
5.0E-06
0.0E+00
0
5
10
15
20
25
30
35
40
45
50
Burnup, MWd/kg
Np-237
BNFL
CEA
ECN
EDF
HIT
IKE1
JAE
PSI1
STU
AVER

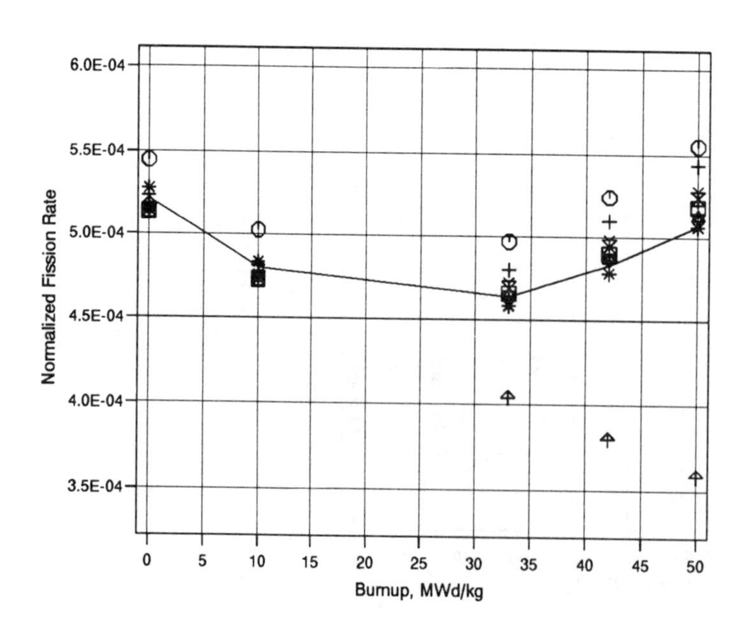

Benchmark B: Fission Rate
Normalized Fission Rate
6.0E-04
5.5E-04
5.0E-04
4.5E-04
4.0E-04
3.5E-04
0
5
10
15
20
25
30
35
40
45
50
Burnup, MWd/kg
Pu-238
BNFL
CEA
ECN
EDF
HIT
IKE1
JAE
PSI1
STU
AVER

Benchmark B: Fission Rate

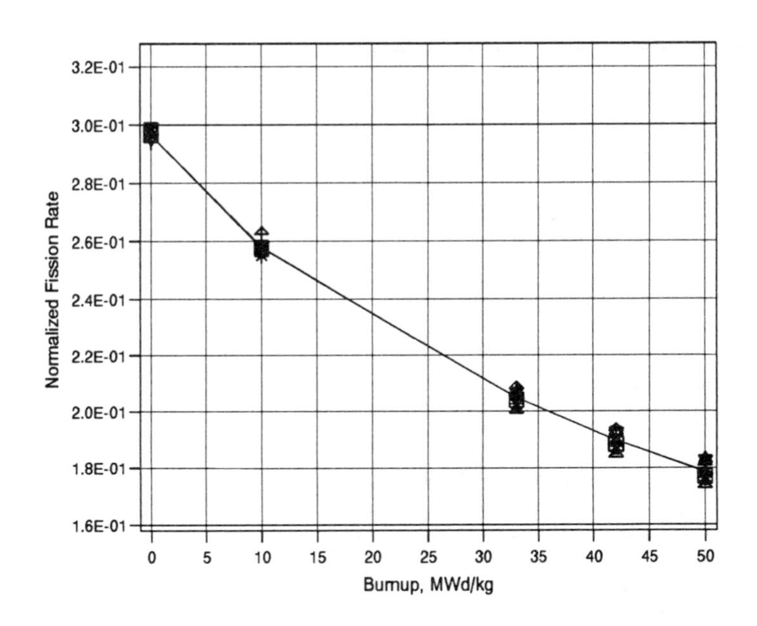

Pu-239

BNFL
CEA
ECN
EDF
HIT
IKE1
JAE
PSI1
STU
AVER

Benchmark B: Fission Rate

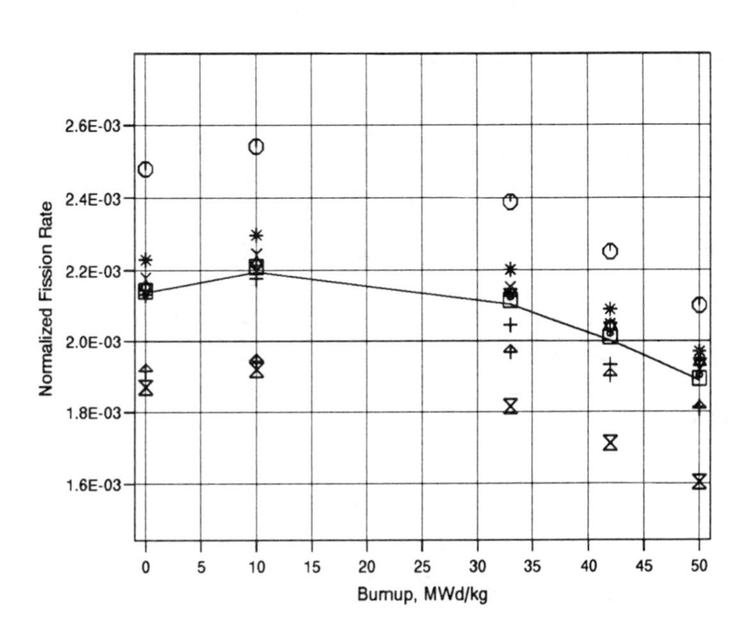

Pu-240

B-fr

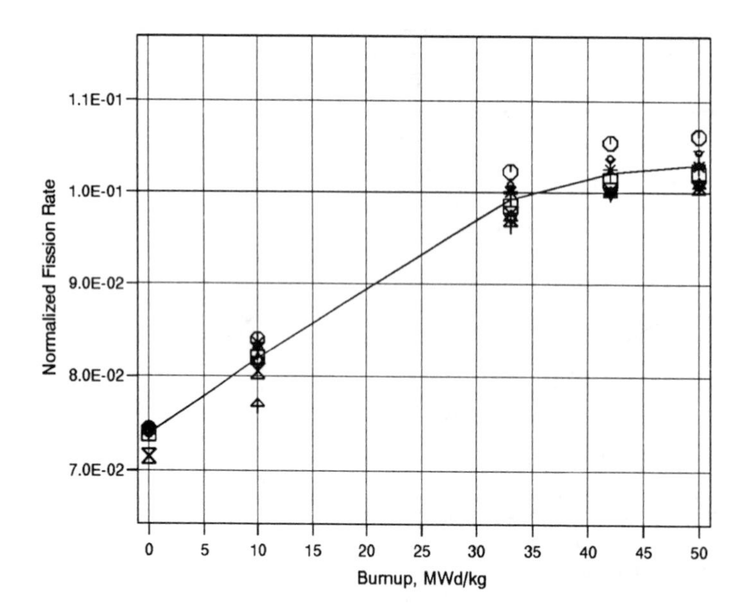

Benchmark B: Fission Rate
Normalized Fission Rate
1.1E-01
1.0E-01
9.0E-02
8.0E-02
7.0E-02
0
5
10
15
20
25
30
35
40
45
50
Burnup, MWd/kg
Pu-241
BNFL
CEA
ECN
EDF
HIT
IKE1
JAE
PSI1
STU
AVER

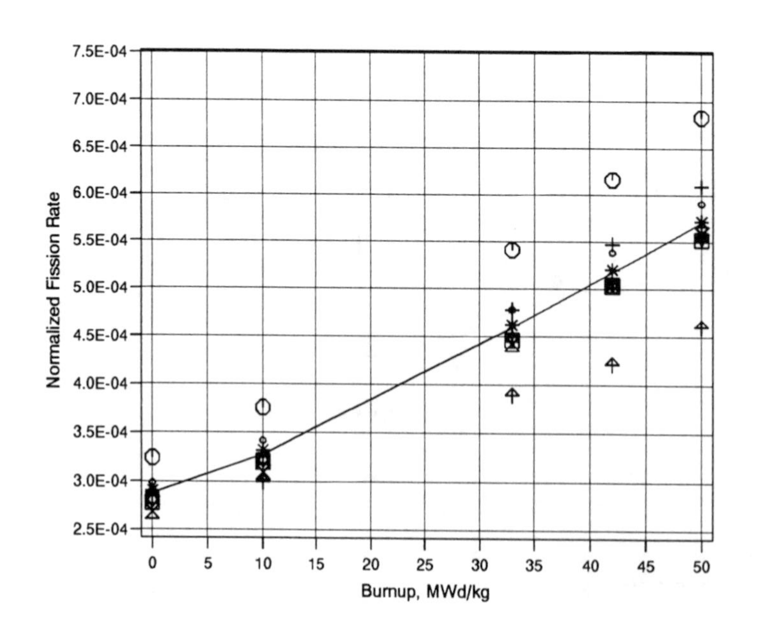

Benchmark B: Fission Rate
Normalized Fission Rate
7.5E-04
7.0E-04
6.5E-04
6.0E-04
5.5E-04
5.0E-04
4.5E-04
4.0E-04
3.5E-04
3.0E-04
2.5E-04
0
5
10
15
20
25
30
35
40
45
50
Burnup, MWd/kg
Pu-242
BNFL
CEA
ECN
EDF
HIT
IKE1
JAE
PSI1
STU
AVER

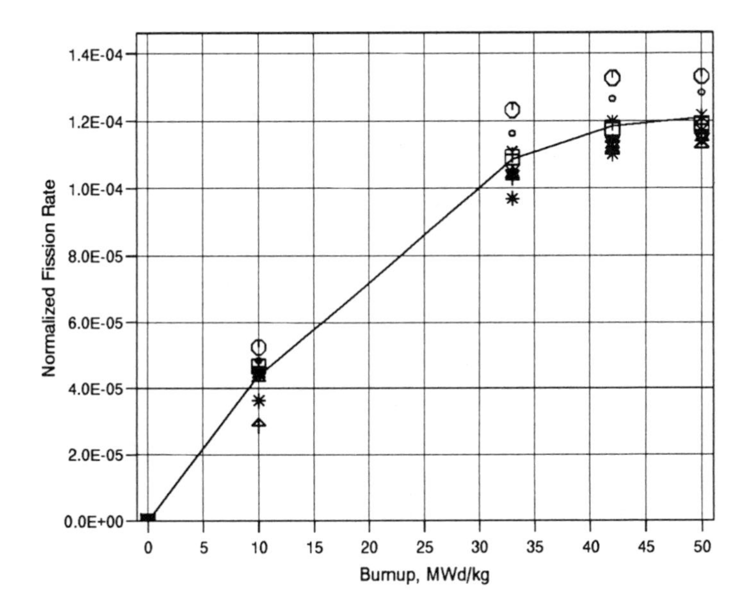

Benchmark B: Fission Rate
Normalized Fission Rate
1.4E-04
1.2E-04
1.0E-04
8.0E-05
6.0E-05
4.0E-05
2.0E-05
0.0E+00
0
5
10
15
20
25
30
35
40
45
50
Burnup, MWd/kg
Am-241
BNFL
CEA
ECN
EDF
HIT
IKE1
JAE
PSI1
STU
AVER

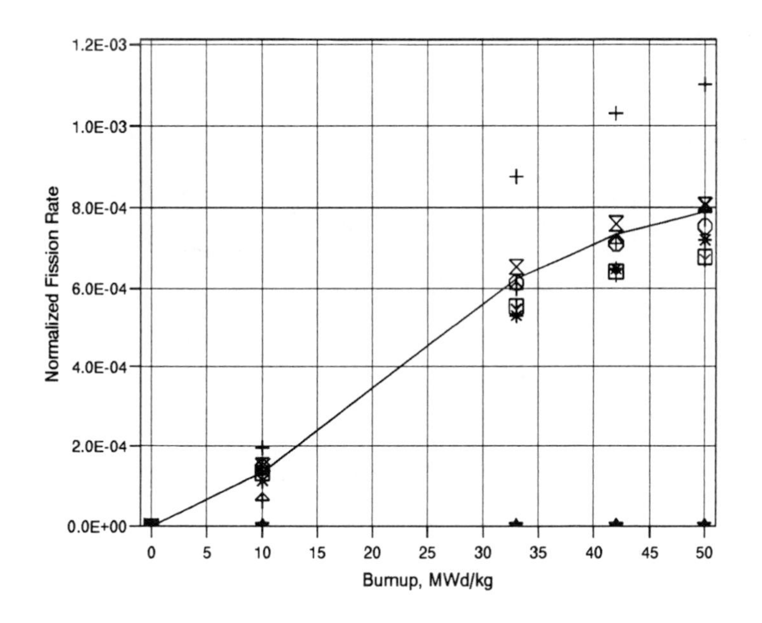

Benchmark B: Fission Rate
Normalized Fission Rate
1.2E-03
1.0E-03
8.0E-04
6.0E-04
4.0E-04
2.0E-04
0.0E+00
0
5
10
15
20
25
30
35
40
45
50
Burnup, MWd/kg
Am-242m
BNFL
CEA
ECN
EDF
HIT
IKE1
JAE
PSI1
STU
AVER

B-fr

Benchmark B: Fission Rate

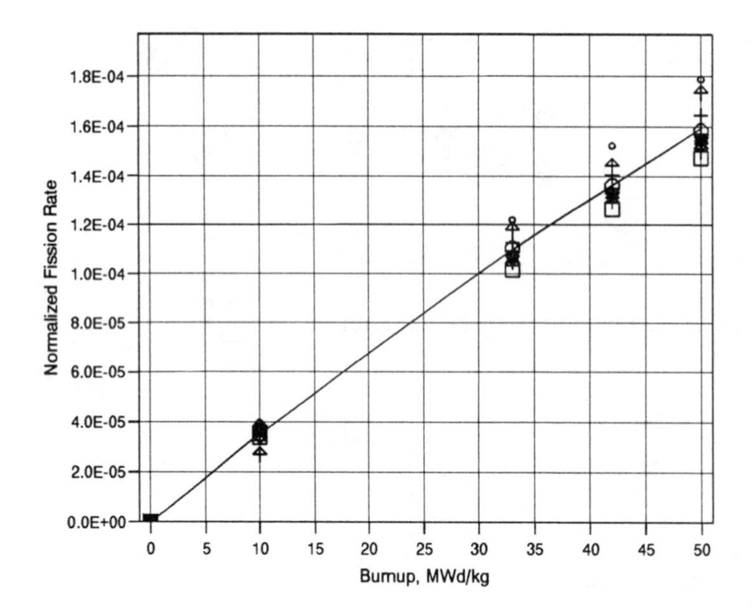

Am-243
BNFL
CEA
ECN
EDF
HIT
IKE1
JAE
PSI1
STU
AVER

Benchmark B: Fission Rate

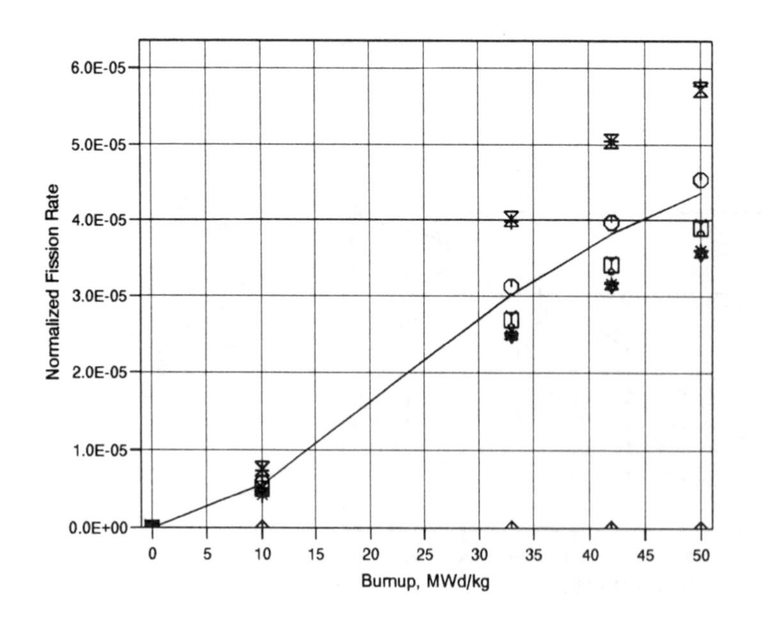

Cm-242

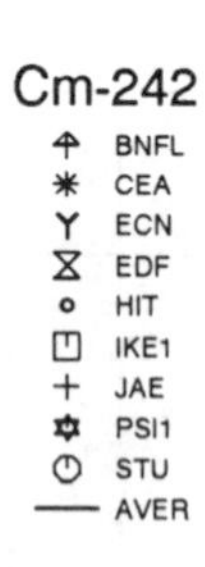

Benchmark B: Fission Rate

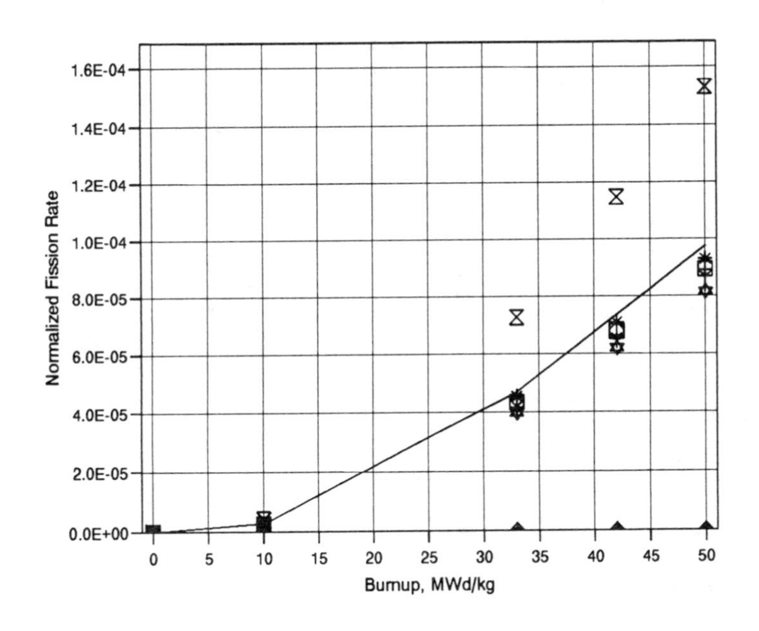

Benchmark B: Fission Rate

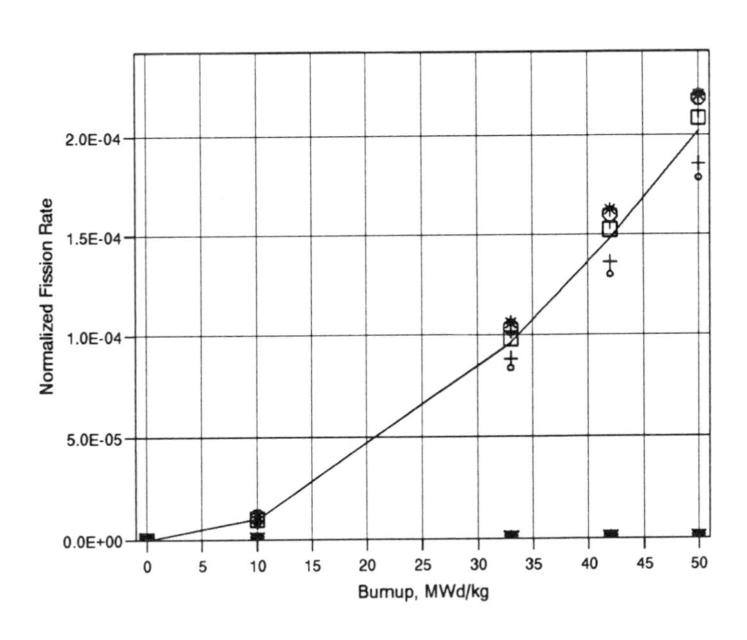

B-fr

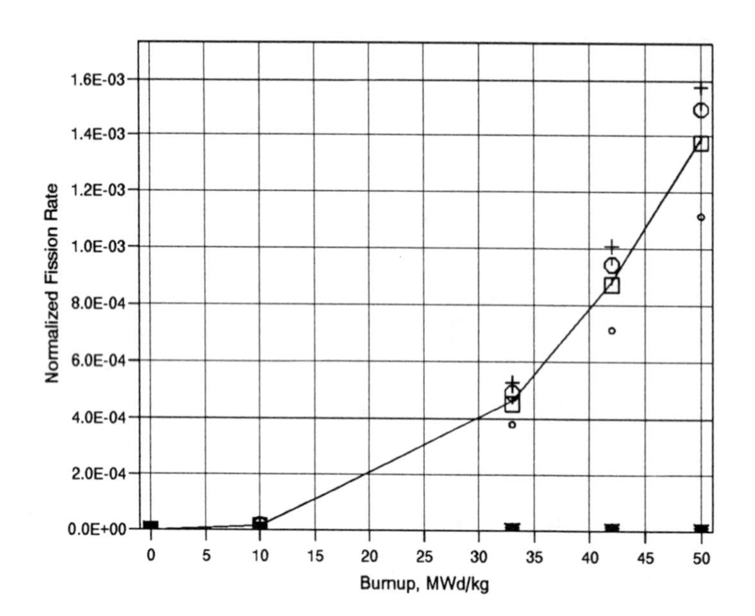

Benchmark B: Fission Rate
Normalized Fission Rate
1.6E-03
1.4E-03
1.2E-03
1.0E-03
8.0E-04
6.0E-04
4.0E-04
2.0E-04
0.0E+00
0
5
10
15
20
25
30
35
40
45
50
Burnup, MWd/kg
Cm-245
BNFL
CEA
ECN
EDF
HIT
IKE1
JAE
PSI1
STU
AVER

Contributor: BNFL

B

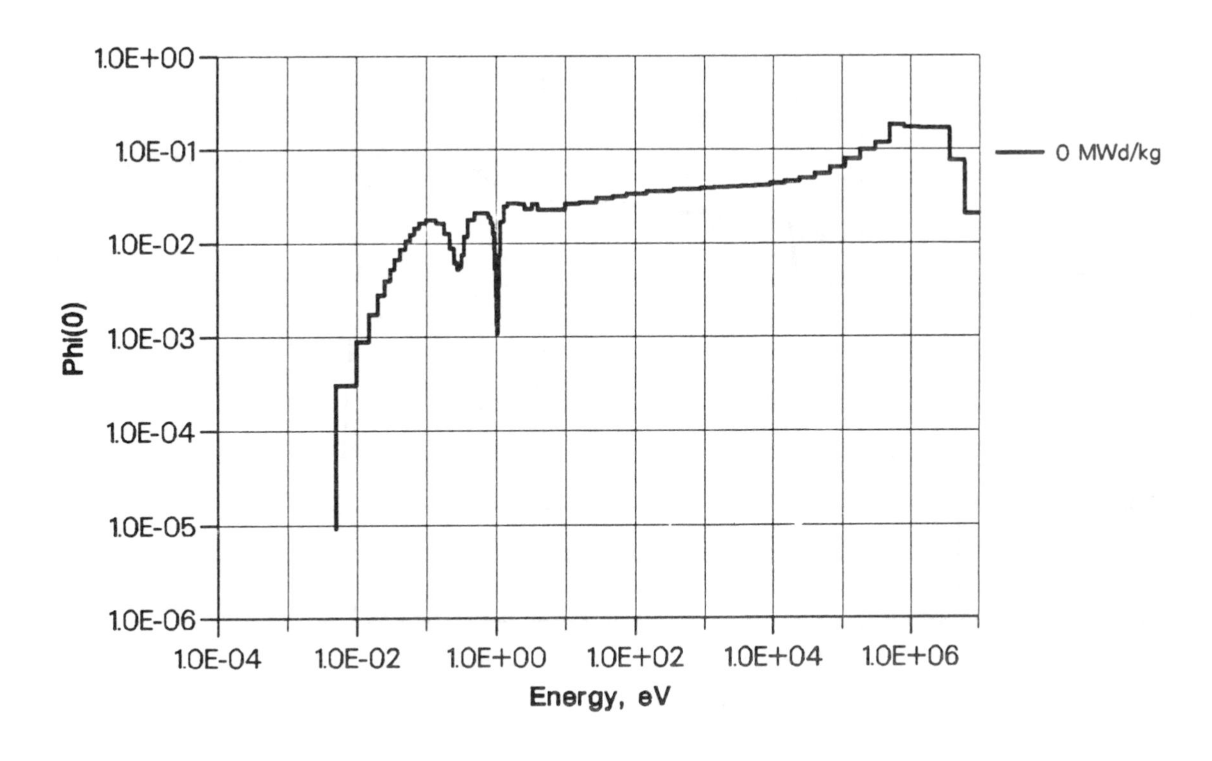

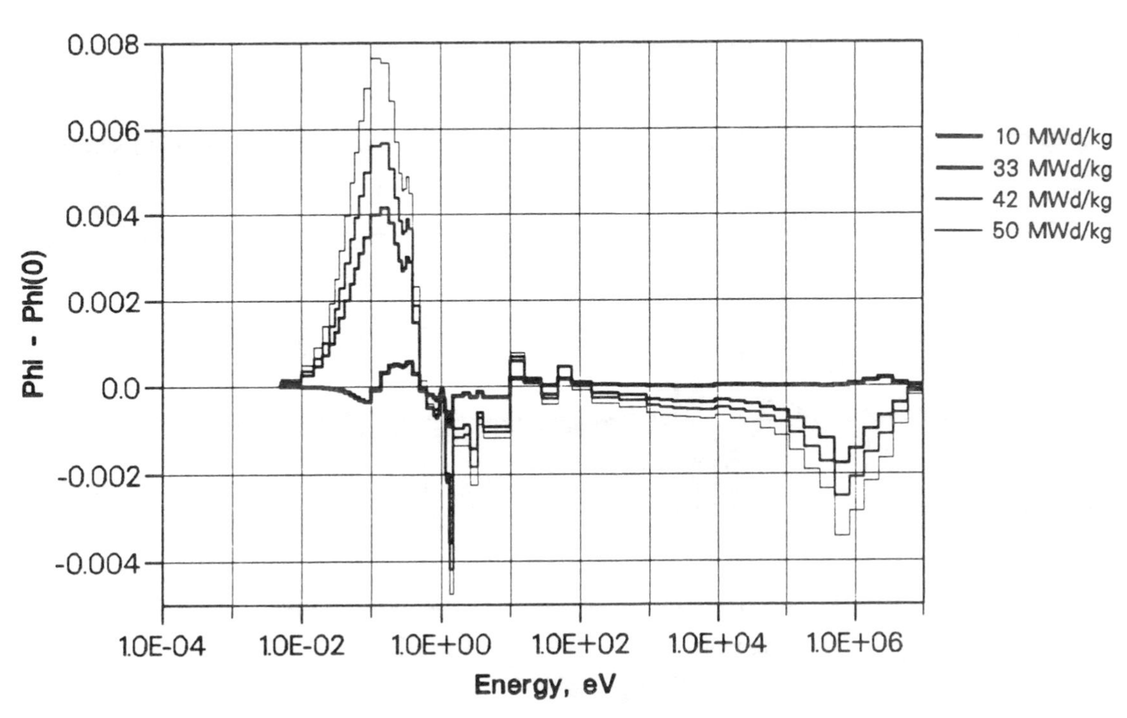

Benchmark 1-B SPECTRA as a function of burnup

B

Contributor: CEA

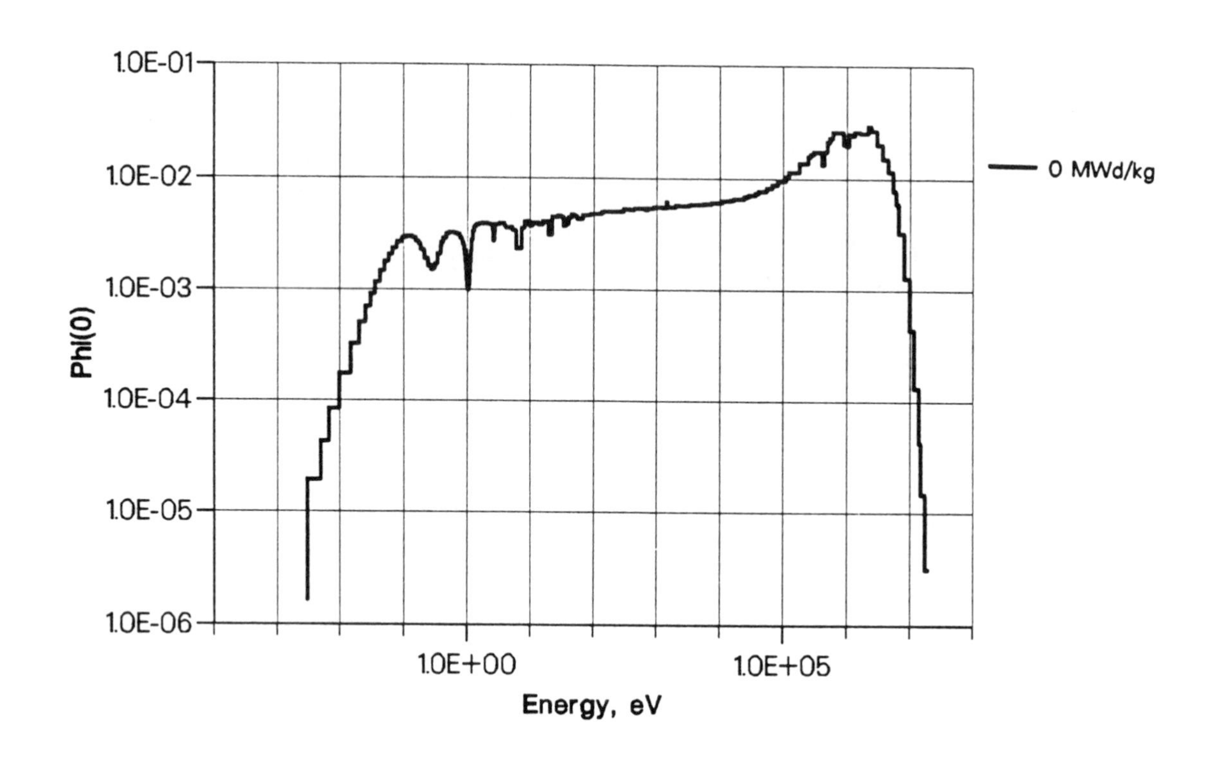

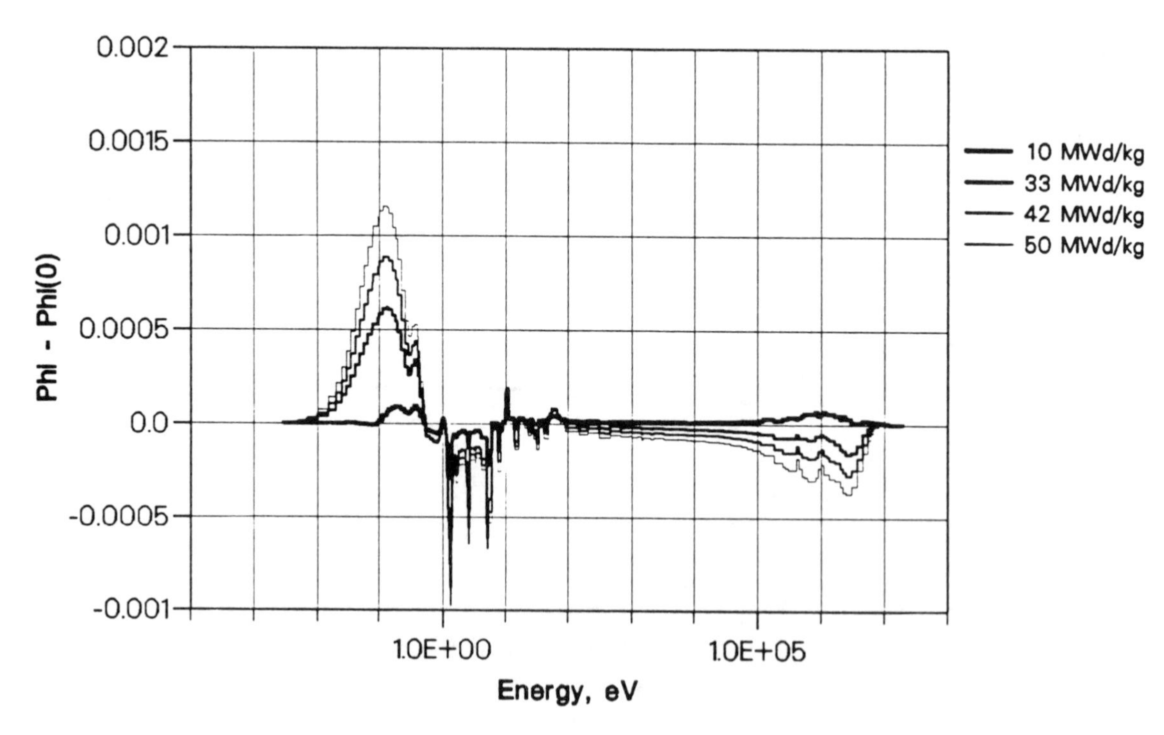

Benchmark 1-B SPECTRA as a function of burnup

Contributor: ECN

B

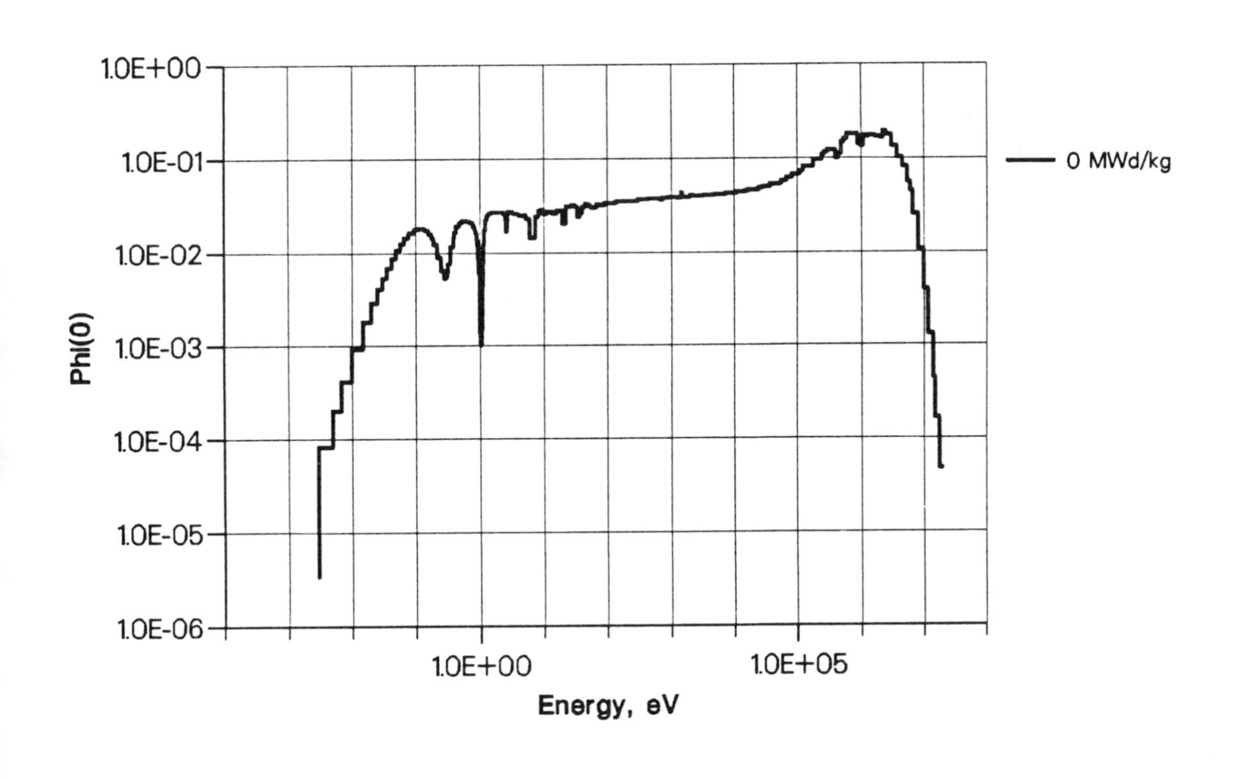

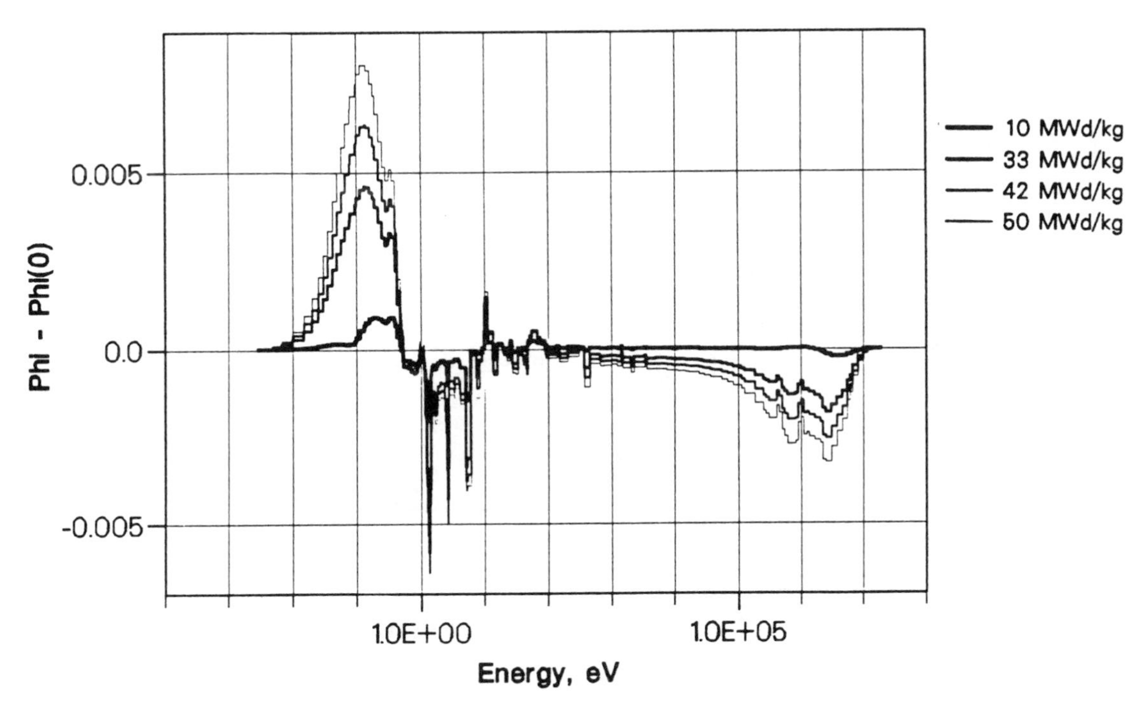

Benchmark 1-B SPECTRA as a function of burnup

Contributor: EDF

B

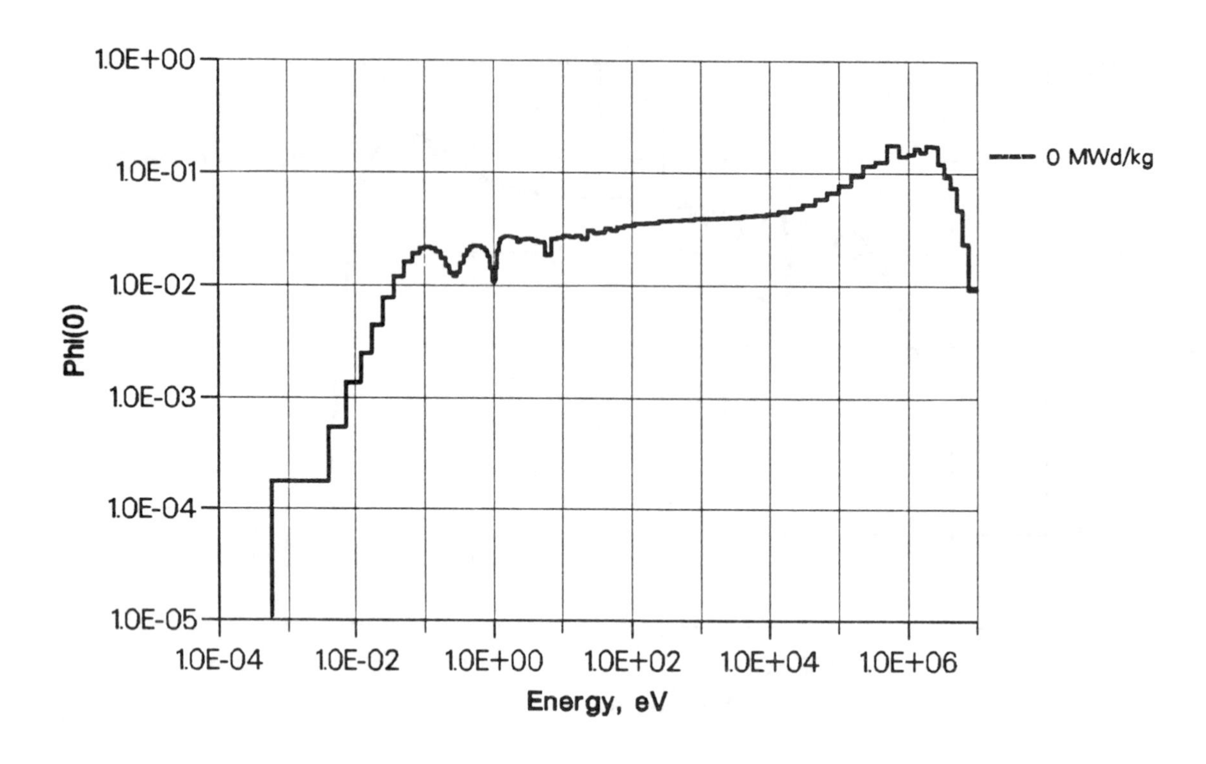

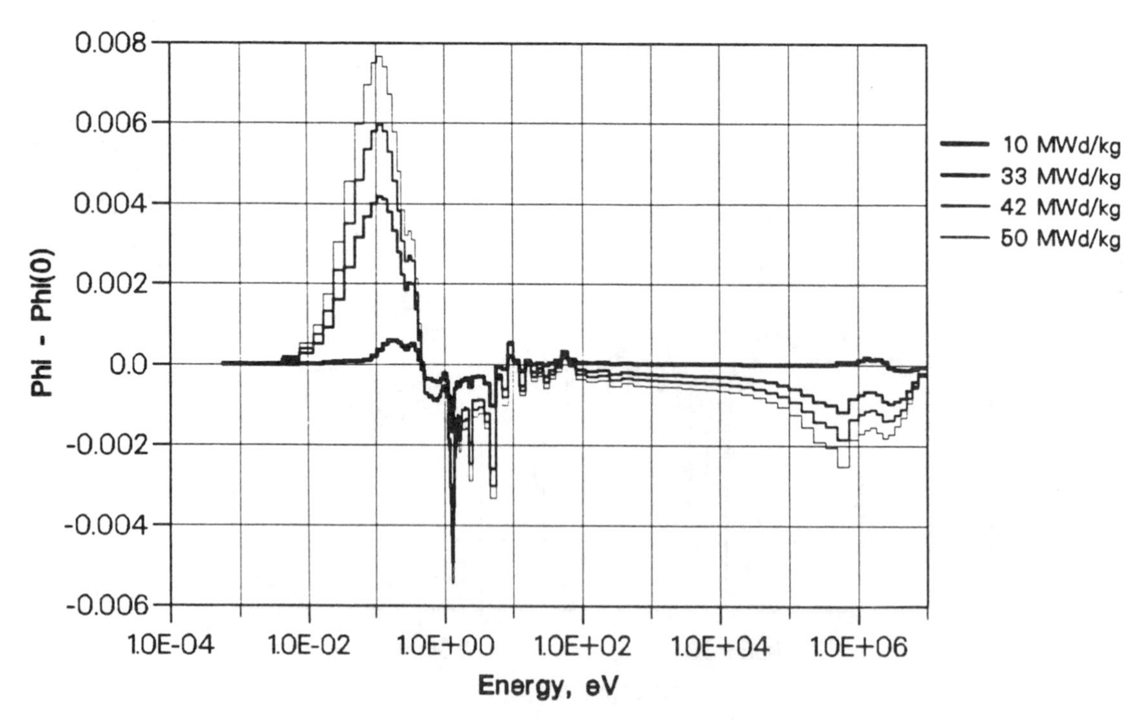

Benchmark 1-B SPECTRA as a function of burnup

Contributor: IKE 1

B

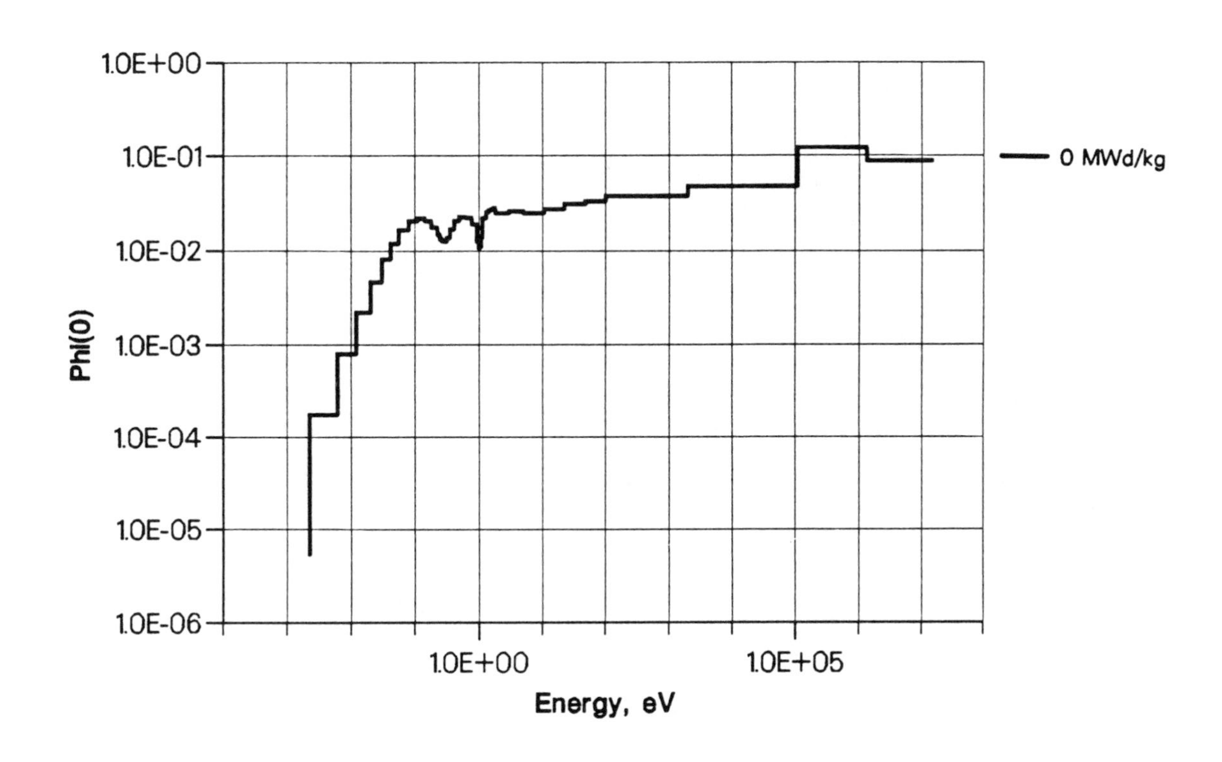

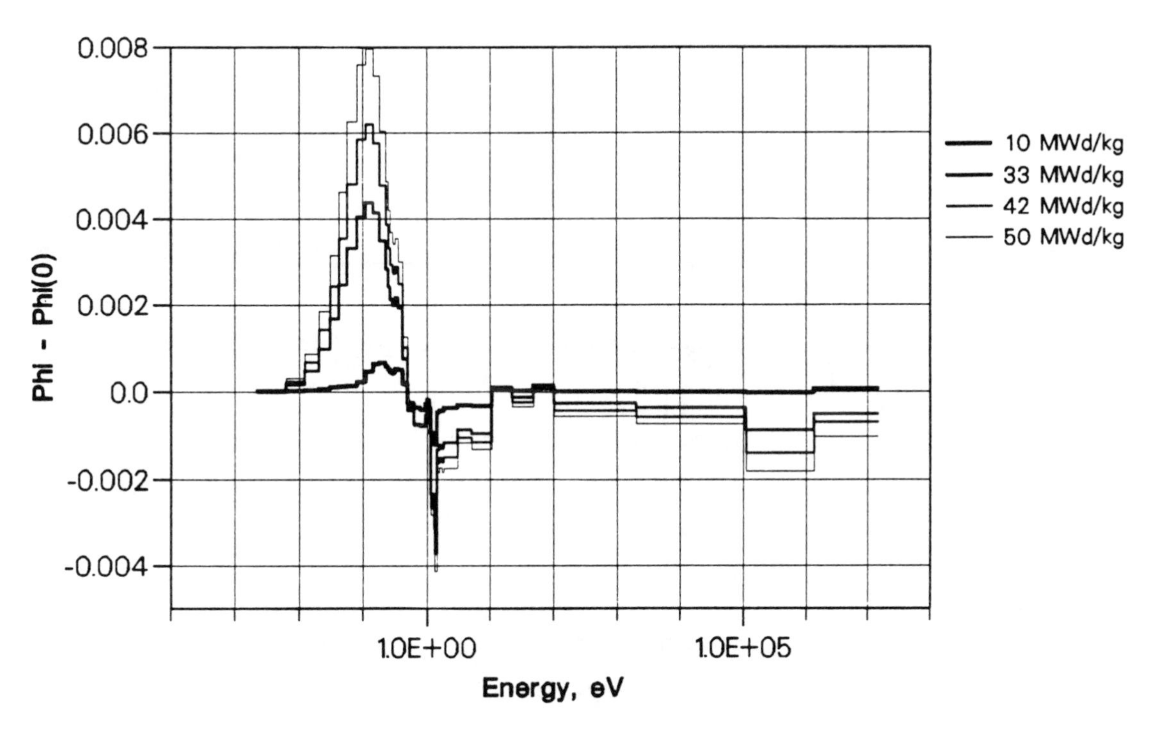

Benchmark 1-B SPECTRA as a function of burnup

B

Contributor: JAERI

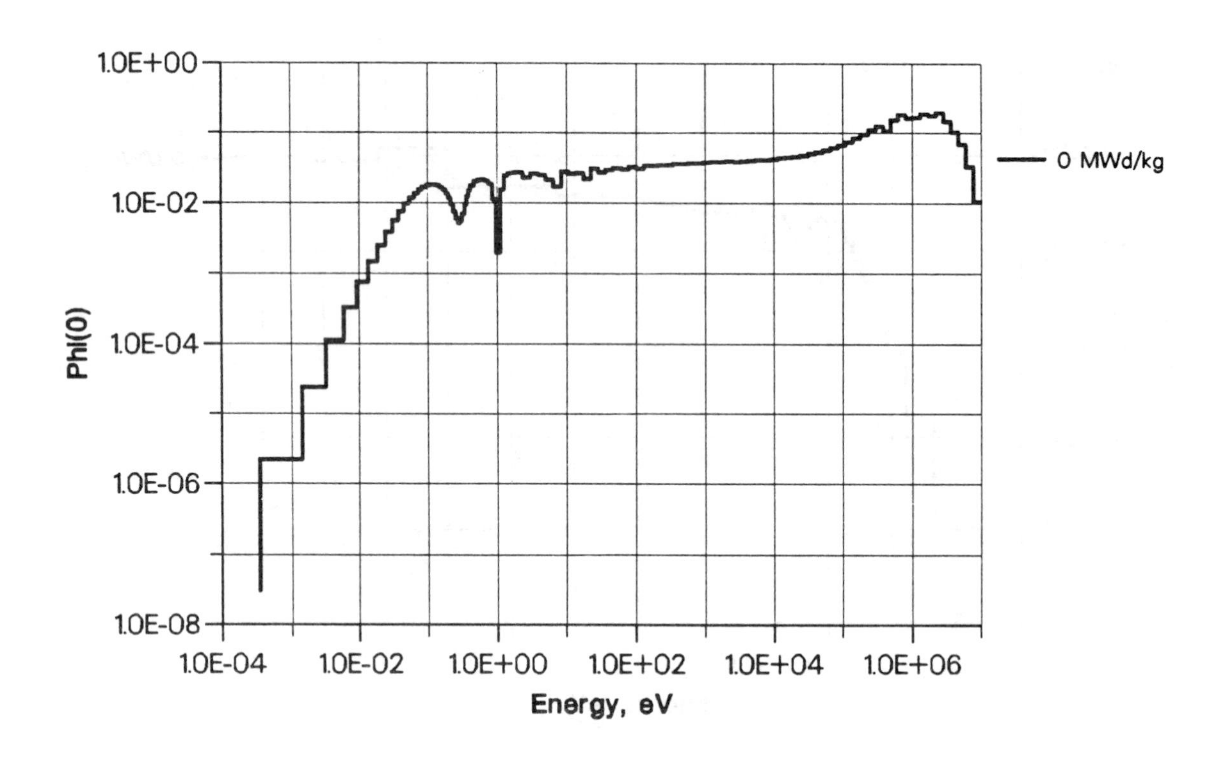

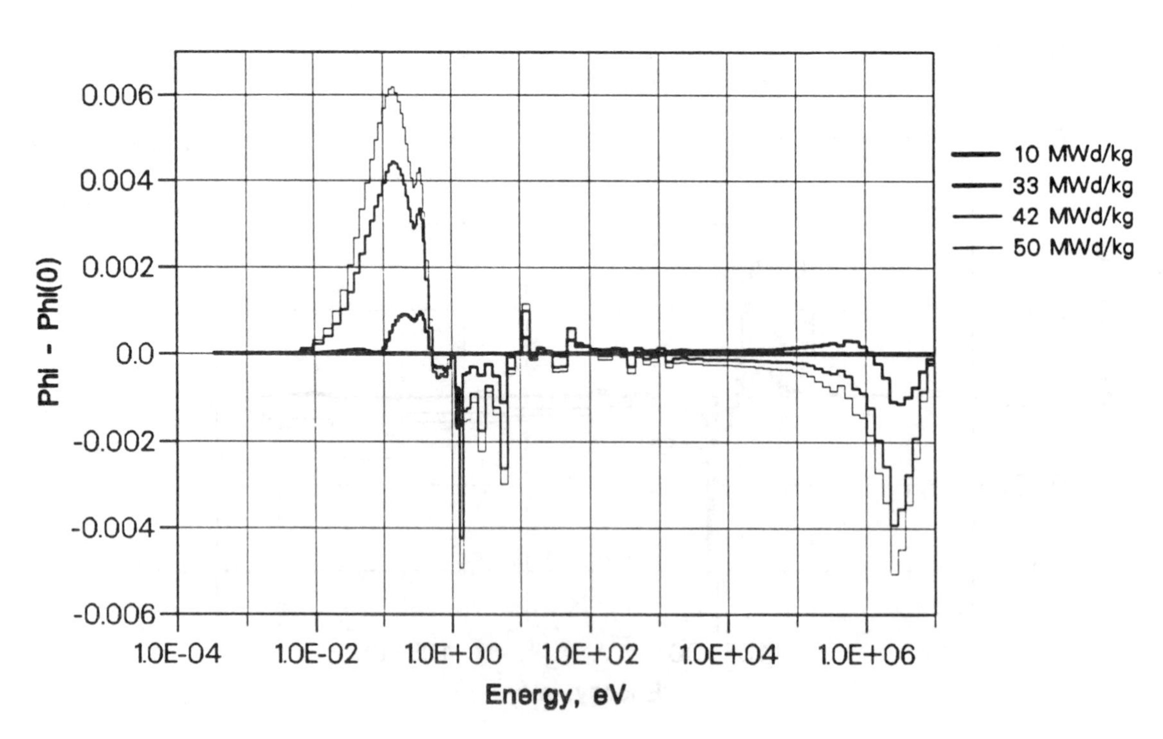

Benchmark 1-B SPECTRA as a function of burnup

Contributor: PSI 1

B

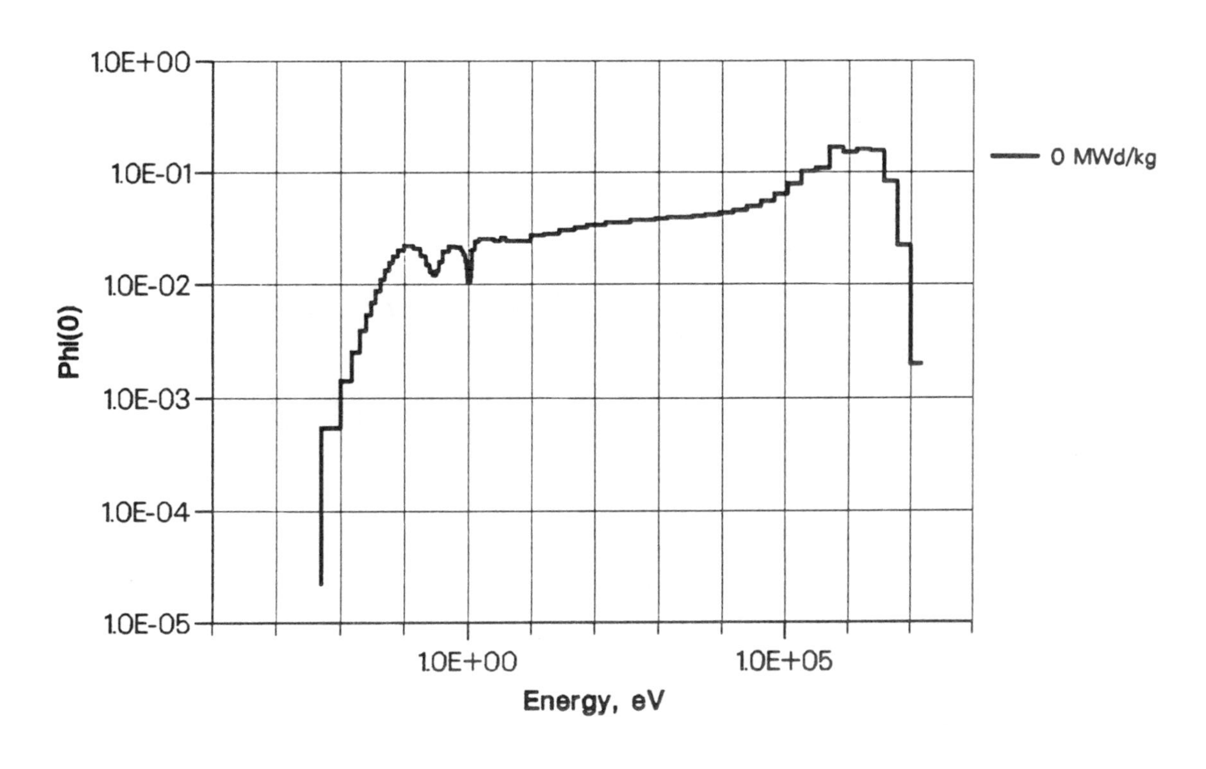

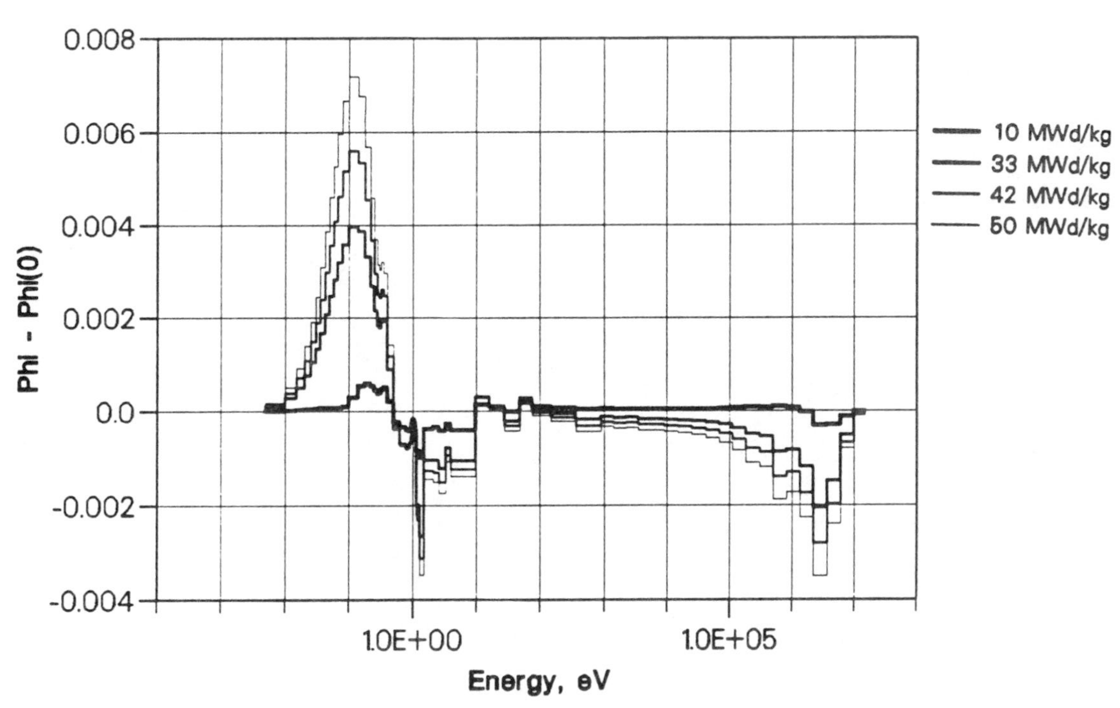

Benchmark 1-B SPECTRA as a function of burnup

Appendix D.1

Sensitivity calculations for benchmark problem A and B

A. Puill and S. Cathalau (CEA)

A number of calculations were performed with APOLLO-2 in order to assess reactivity worths linked to the differences in data and/or models used by the participants. The corresponding results are given hereafter.

Temperature correction for IKE-2 solution

Two APOLLO-2 calculations were performed at IKE-2 temperatures; the results are the following:

	SPECIFIED TEMPERATURES (660, 306.3, 306.3 °C)	IKE-2 TEMPERATURES (626.85, 326.85, 300.45°C)	Δk/k (pcm)
k-infinity – case A	1.13341	1.13460	-93
k-infinity – case B	1.18947	1.19067	-85

Those corrections have been added to the original IKE-2 results.

Zirconium correction

This correction was calculated only for fresh fuel conditions. A simplified self-shielding model was used (no spatial discretisation of the fuel pin). The difference between natural Zr and Zr-91 is very large because Zr-91 is the most absorbing isotope of zirconium, but has an isotopic abundance of only 12%.

	NATURAL Zr	Zr-91	Δk/k (pcm)
k-infinity – case A	1.12966	1.11364	-1273
k-infinity – case B	1.18515	1.16348	-1572

Correction linked with the number of isotopes contributing to energy release

The isotopes specified as contributing to energy release are: U-235, U-238, Pu-239, Pu-241, Am-242. In the standard depletion chain of APOLLO-2, the following 18 isotopes are taken into account: U-234 to U-236, U-238, Np-237 to Np-239, Pu-238 to Pu-242, Am-241 to Am-243, Cm-242 to Cm-244. The number of fission products is 77. The difference in reactivity resulting from using more isotopes than specified is the following:

		SPECIFIED NUMBER	APOLLO-2 STANDARD NUMBER	Δk/k (pcm)
Case A	0 MWd/kg	1.13341	1.13349	+6
Case A	50 MWd/kg	0.94970	0.95325	+392
Case B	0 MWd/kg	1.18947	1.18953	+4
Case B	50 MWd/kg	0.91758	0.91923	+196

The effects shown in the above table do not compensate for the differences observed in the reactivity loss during irradiation, even when solutions derived from the same evaluated data file are considered: for example, the reactivity loss derived from JEF-2.2 ranges from 15500 to 17100 pcm (in Δk/k) for *benchmark problem A*, and from 22900 to 24900 pcm for *benchmark problem B*. Differences between fission yields used could explain the remaining spreads.

Self-shielding model corrections

Two corrections were assessed: the first one is connected with the fuel pin spatial discretisation which allows to take into account the variation of the shielding factors within the pin; the second one deals with the variation of those factors during irradiation. Those corrections apply only to deterministic codes.

Spatial discretisation effect

		1 MESH IN THE PIN	6 MESHES IN THE PIN	Δk/k (pcm)
Case A	0 MWd/kg	1.12966	1.13341	+293
Case A	50 MWd/kg	0.94750	0.94970	+244
Case B	0 MWd/kg	1.18515	1.18947	+306
Case B	50 MWd/kg	0.91779	0.91758	-25

Burnup effect

In the APOLLO-2 solution, shielded cross-sections were calculated at six burnup values: 0, 10, 22, 33, 42 and 50 MWd/kg in order to take into account the changes in the nuclide concentrations during irradiation. The effect of calculating shielded cross-sections only once is small, as shown in the table below.

		1 BURNUP POINT	6 BURNUP POINTS	Δk/k (pcm)
Case A	0 MWd/kg	1.13341	1.13341	0
Case A	50 MWd/kg	0.95049	0.94970	-88
Case B	0 MWd/kg	1.18947	1.18497	+306
Case B	50 MWd/kg	0.91810	0.91758	-62

Appendix D.2

Comparison of the results calculated with several Monte Carlo codes and nuclear data for plutonium recycling and void coefficient benchmark in PWRs

H. Takano, T. Mori and H. Akie (JAERI)

Comparison of the results calculated with the deterministic code SRAC, and continuous-energy Monte Carlo code MVP for the poor-quality plutonium cell model

Results calculated for a temperature of 300 K and 600 K are in very good agreement with each other.

The difference in the case of $T = 900$ K is due to the Doppler effect for plutonium isotopes.

	$T_f = 900$ K	600 K	300 K
SRAC	1.1347	1.1464	1.1626
MVP	1.1372 ± 0.07% [1]	1.1457 ± 0.08% [2]	1.1631 ± 0.07%

Infinite multiplication factors calculated for poor-quality plutonium cell with different fuel temperatures by using JENDL-3.1 nuclear data

[1] $T = 900$ K for U-235 and U-238,
600 K for Pu-239 - Pu-242, and
300 K for Pu-238.

[2] $T = 600$ K for U-235, U-238 and Pu-239 - Pu-242, and
300 K for Pu-238.

Appendix D.3

Benchmark calculations for plutonium recycling in PWRs

E. Sajii (Toden Software Ltd.)

Calculation code **CASMO-4**

Nuclear data library **JEF-2.2 and ENDF/B-IV**

Benchmark A Poor-quality plutonium

Burnup (MWd/kg)	JEF-2.2	ENDF/B-VI
0.0	1.1321	1.1337
10	1.0731	1.0749
33	1.0037	1.0038
42	0.9809	0.9788
50	0.9624	0.9576

Benchmark B Better-quality plutonium

Burnup (MWd/kg)	JEF-2.2	ENDF/B-VI
0.0	1.1810	1.1806
10	1.0931	1.0930
33	0.9897	0.9885
42	0.9551	0.9522
50	0.9263	0.9213

List of symbols and abbreviations

at/cc	atoms /cm^3
atom/cm^3	atomic density
AR	*A*bsorption *R*ate
BOL	*B*eginning *O*f *L*ife
BWR	*B*oiling-*W*ater *R*eactor
°C	degrees Celsius
CPU	*C*entral *P*rocessor *U*nit
dAR	variation of absorption rates
dFR	variation in normalized fission rates
Δk	variation of multiplication factor
eV	electronvolt
FP	*F*ission *P*roduct
FR	*F*ission *R*ate
g/cm^3	mass density
g/mol	atomic mass
GWd/t	gigawattdays per ton (metal) = **MWd/kg**
ID	nuclide identifier
k	neutron multiplication factor
K	degrees Kelvin
keV	kiloelectronvolt
LWR	*L*ight-*W*ater *R*eactor
MeV	megaelectronvolt
MOX	*M*ixed *O*xide (uranium and plutonium)
MW	megawatt
MWd/kg	megawattdays per kilogram (metal)

ν (nu)	neutron per fission
NSC	*N*uclear *S*cience *C*ommittee
OECD/NEA	OECD *N*uclear *E*nergy *A*gency
pcm	10^{-5}
ppm	parts per million
Pu	plutonium
PWR	*P*ressurized-*W*ater *R*eactor
σ_p	potential cross-section
$S(\alpha,\beta)$	thermal scattering law
S_n	discrete ordinates radiation transport method
T	temperature
U	uranium
UO_2	uranium dioxide = **UOX**
W/g	Watt per gram
w/o	weight % = **wt%**
WPPR	*W*orking Party on the *P*hysics of *P*lutonium *R*ecycling

MAIN SALES OUTLETS OF OECD PUBLICATIONS
PRINCIPAUX POINTS DE VENTE DES PUBLICATIONS DE L'OCDE

ARGENTINA – ARGENTINE
Carlos Hirsch S.R.L.
Galería Güemes, Florida 165, 4° Piso
1333 Buenos Aires Tel. (1) 331.1787 y 331.2391
Telefax: (1) 331.1787

AUSTRALIA – AUSTRALIE
D.A. Information Services
648 Whitehorse Road, P.O.B 163
Mitcham, Victoria 3132 Tel. (03) 873.4411
Telefax: (03) 873.5679

AUSTRIA – AUTRICHE
Gerold & Co.
Graben 31
Wien I Tel. (0222) 533.50.14
Telefax: (0222) 512.47.31.29

BELGIUM – BELGIQUE
Jean De Lannoy
Avenue du Roi 202 Koningslaan
B-1060 Bruxelles Tel. (02) 538.51.69/538.08.41
Telefax: (02) 538.08.41

CANADA
Renouf Publishing Company Ltd.
1294 Algoma Road
Ottawa, ON K1B 3W8 Tel. (613) 741.4333
Telefax: (613) 741.5439

Stores:
61 Sparks Street
Ottawa, ON K1P 5R1 Tel. (613) 238.8985
211 Yonge Street
Toronto, ON M5B 1M4 Tel. (416) 363.3171
Telefax: (416)363.59.63

Les Éditions La Liberté Inc.
3020 Chemin Sainte-Foy
Sainte-Foy, PQ G1X 3V6 Tel. (418) 658.3763
Telefax: (418) 658.3763

Federal Publications Inc.
165 University Avenue, Suite 701
Toronto, ON M5H 3B8 Tel. (416) 860.1611
Telefax: (416) 860.1608

Les Publications Fédérales
1185 Université
Montréal, QC H3B 3A7 Tel. (514) 954.1633
Telefax: (514) 954.1635

CHINA – CHINE
China National Publications Import
Export Corporation (CNPIEC)
16 Gongti E. Road, Chaoyang District
P.O. Box 88 or 50
Beijing 100704 PR Tel. (01) 506.6688
Telefax: (01) 506.3101

CHINESE TAIPEI – TAIPEI CHINOIS
Good Faith Worldwide Int'l. Co. Ltd.
9th Floor, No. 118, Sec. 2
Chung Hsiao E. Road
Taipei Tel. (02) 391.7396/391.7397
Telefax: (02) 394.9176

CZECH REPUBLIC – RÉPUBLIQUE TCHEQUE
Artia Pegas Press Ltd.
Narodni Trida 25
POB 825
111 21 Praha 1 Tel. 26.65.68
Telefax: 26.20.81

DENMARK – DANEMARK
Munksgaard Book and Subscription Service
35, Nørre Søgade, P.O. Box 2148
DK-1016 København K Tel. (33) 12.85.70
Telefax: (33) 12.93.87

EGYPT – ÉGYPTE
Middle East Observer
41 Sherif Street
Cairo Tel. 392.6919
Telefax: 360-6804

FINLAND – FINLANDE
Akateeminen Kirjakauppa
Keskuskatu 1, P.O. Box 128
00100 Helsinki

Subscription Services/Agence d'abonnements :
P.O. Box 23
00371 Helsinki Tel. (358 0) 121 4416
Telefax: (358 0) 121.4450

FRANCE
OECD/OCDE
Mail Orders/Commandes par correspondance:
2, rue André-Pascal
75775 Paris Cedex 16 Tel. (33-1) 45.24.82.00
Telefax: (33-1) 49.10.42.76
Telex: 640048 OCDE

Internet: Compte.PUBSINQ @ oecd.org

Orders via Minitel, France only/
Commandes par Minitel, France exclusivement :
36 15 OCDE

OECD Bookshop/Librairie de l'OCDE :
33, rue Octave-Feuillet
75016 Paris Tel. (33-1) 45.24.81.81
(33-1) 45.24.81.67

Documentation Française
29, quai Voltaire
75007 Paris Tel. 40.15.70.00

Gibert Jeune (Droit-Économie)
6, place Saint-Michel
75006 Paris Tel. 43.25.91.19

Librairie du Commerce International
10, avenue d'Iéna
75016 Paris Tel. 40.73.34.60

Librairie Dunod
Université Paris-Dauphine
Place du Maréchal de Lattre de Tassigny
75016 Paris Tel. (1) 44.05.40.13

Librairie Lavoisier
11, rue Lavoisier
75008 Paris Tel. 42.65.39.95

Librairie L.G.D.J. - Montchrestien
20, rue Soufflot
75005 Paris Tel. 46.33.89.85

Librairie des Sciences Politiques
30, rue Saint-Guillaume
75007 Paris Tel. 45.48.36.02

P.U.F.
49, boulevard Saint-Michel
75005 Paris Tel. 43.25.83.40

Librairie de l'Université
12a, rue Nazareth
13100 Aix-en-Provence Tel. (16) 42.26.18.08

Documentation Française
165, rue Garibaldi
69003 Lyon Tel. (16) 78.63.32.23

Librairie Decitre
29, place Bellecour
69002 Lyon Tel. (16) 72.40.54.54

Librairie Sauramps
Le Triangle
34967 Montpellier Cedex 2 Tel. (16) 67.58.85.15
Tekefax: (16) 67.58.27.36

GERMANY – ALLEMAGNE
OECD Publications and Information Centre
August-Bebel-Allee 6
D-53175 Bonn Tel. (0228) 959.120
Telefax: (0228) 959.12.17

GREECE – GRÈCE
Librairie Kauffmann
Mavrokordatou 9
106 78 Athens Tel. (01) 32.55.321
Telefax: (01) 32.30.320

HONG-KONG
Swindon Book Co. Ltd.
Astoria Bldg. 3F
34 Ashley Road, Tsimshatsui
Kowloon, Hong Kong Tel. 2376.2062
Telefax: 2376.0685

HUNGARY – HONGRIE
Euro Info Service
Margitsziget, Európa Ház
1138 Budapest Tel. (1) 111.62.16
Telefax: (1) 111.60.61

ICELAND – ISLANDE
Mál Mog Menning
Laugavegi 18, Pósthólf 392
121 Reykjavik Tel. (1) 552.4240
Telefax: (1) 562.3523

INDIA – INDE
Oxford Book and Stationery Co.
Scindia House
New Delhi 110001 Tel. (11) 331.5896/5308
Telefax: (11) 332.5993

17 Park Street
Calcutta 700016 Tel. 240832

INDONESIA – INDONÉSIE
Pdii-Lipi
P.O. Box 4298
Jakarta 12042 Tel. (21) 573.34.67
Telefax: (21) 573.34.67

IRELAND – IRLANDE
Government Supplies Agency
Publications Section
4/5 Harcourt Road
Dublin 2 Tel. 661.31.11
Telefax: 475.27.60

ISRAEL
Praedicta
5 Shatner Street
P.O. Box 34030
Jerusalem 91430 Tel. (2) 52.84.90/1/2
Telefax: (2) 52.84.93

R.O.Y. International
P.O. Box 13056
Tel Aviv 61130 Tel. (3) 546 1423
Telefax: (3) 546 1442

Palestinian Authority/Middle East:
INDEX Information Services
P.O.B. 19502
Jerusalem Tel. (2) 27.12.19
Telefax: (2) 27.16.34

ITALY – ITALIE
Libreria Commissionaria Sansoni
Via Duca di Calabria 1/1
50125 Firenze Tel. (055) 64.54.15
Telefax: (055) 64.12.57

Via Bartolini 29
20155 Milano Tel. (02) 36.50.83

Editrice e Libreria Herder
Piazza Montecitorio 120
00186 Roma Tel. 679.46.28
Telefax: 678.47.51

Libreria Hoepli
Via Hoepli 5
20121 Milano Tel. (02) 86.54.46
Telefax: (02) 805.28.86

Libreria Scientifica
Dott. Lucio de Biasio 'Aeiou'
Via Coronelli, 6
20146 Milano Tel. (02) 48.95.45.52
Telefax: (02) 48.95.45.48

JAPAN – JAPON
OECD Publications and Information Centre
Landic Akasaka Building
2-3-4 Akasaka, Minato-ku
Tokyo 107 Tel. (81.3) 3586.2016
Telefax: (81.3) 3584.7929

KOREA – CORÉE
Kyobo Book Centre Co. Ltd.
P.O. Box 1658, Kwang Hwa Moon
Seoul Tel. 730.78.91
Telefax: 735.00.30

MALAYSIA – MALAISIE
University of Malaya Bookshop
University of Malaya
P.O. Box 1127, Jalan Pantai Baru
59700 Kuala Lumpur
Malaysia Tel. 756.5000/756.5425
Telefax: 756.3246

MEXICO – MEXIQUE
Revistas y Periodicos Internacionales S.A. de C.V.
Florencia 57 - 1004
Mexico, D.F. 06600 Tel. 207.81.00
Telefax: 208.39.79

NETHERLANDS – PAYS-BAS
SDU Uitgeverij Plantijnstraat
Externe Fondsen
Postbus 20014
2500 EA's-Gravenhage Tel. (070) 37.89.880
Voor bestellingen: Telefax: (070) 34.75.778

NEW ZEALAND
NOUVELLE-ZÉLANDE
GPLegislation Services
P.O. Box 12418
Thorndon, Wellington Tel. (04) 496.5655
Telefax: (04) 496.5698

NORWAY – NORVÈGE
Narvesen Info Center – NIC
Bertrand Narvesens vei 2
P.O. Box 6125 Etterstad
0602 Oslo 6 Tel. (022) 57.33.00
Telefax: (022) 68.19.01

PAKISTAN
Mirza Book Agency
65 Shahrah Quaid-E-Azam
Lahore 54000 Tel. (42) 353.601
Telefax: (42) 231.730

PHILIPPINE – PHILIPPINES
International Book Center
5th Floor, Filipinas Life Bldg.
Ayala Avenue
Metro Manila Tel. 81.96.76
Telex 23312 RHP PH

PORTUGAL
Livraria Portugal
Rua do Carmo 70-74
Apart. 2681
1200 Lisboa Tel. (01) 347.49.82/5
Telefax: (01) 347.02.64

SINGAPORE – SINGAPOUR
Gower Asia Pacific Pte Ltd.
Golden Wheel Building
41, Kallang Pudding Road, No. 04-03
Singapore 1334 Tel. 741.5166
Telefax: 742.9356

SPAIN – ESPAGNE
Mundi-Prensa Libros S.A.
Castelló 37, Apartado 1223
Madrid 28001 Tel. (91) 431.33.99
Telefax: (91) 575.39.98

Libreria Internacional AEDOS
Consejo de Ciento 391
08009 – Barcelona Tel. (93) 488.30.09
Telefax: (93) 487.76.59

Llibreria de la Generalitat
Palau Moja
Rambla dels Estudis, 118
08002 – Barcelona
(Subscripcions) Tel. (93) 318.80.12
(Publicacions) Tel. (93) 302.67.23
Telefax: (93) 412.18.54

SRI LANKA
Centre for Policy Research
c/o Colombo Agencies Ltd.
No. 300-304, Galle Road
Colombo 3 Tel. (1) 574240, 573551-2
Telefax: (1) 575394, 510711

SWEDEN – SUÈDE
Fritzes Customer Service
S–106 47 Stockholm Tel. (08) 690.90.90
Telefax: (08) 20.50.21

Subscription Agency/Agence d'abonnements :
Wennergren-Williams Info AB
P.O. Box 1305
171 25 Solna Tel. (08) 705.97.50
Telefax: (08) 27.00.71

SWITZERLAND – SUISSE
Maditec S.A. (Books and Periodicals - Livres et périodiques)
Chemin des Palettes 4
Case postale 266
1020 Renens VD 1 Tel. (021) 635.08.65
Telefax: (021) 635.07.80

Librairie Payot S.A.
4, place Pépinet
CP 3212
1002 Lausanne Tel. (021) 341.33.47
Telefax: (021) 341.33.45

Librairie Unilivres
6, rue de Candolle
1205 Genève Tel. (022) 320.26.23
Telefax: (022) 329.73.18

Subscription Agency/Agence d'abonnements :
Dynapresse Marketing S.A.
38 avenue Vibert
1227 Carouge Tel. (022) 308.07.89
Telefax: (022) 308.07.99

See also – Voir aussi :
OECD Publications and Information Centre
August-Bebel-Allee 6
D-53175 Bonn (Germany) Tel. (0228) 959.120
Telefax: (0228) 959.12.17

THAILAND – THAÏLANDE
Suksit Siam Co. Ltd.
113, 115 Fuang Nakhon Rd.
Opp. Wat Rajbopith
Bangkok 10200 Tel. (662) 225.9531/2
Telefax: (662) 222.5188

TURKEY – TURQUIE
Kültür Yayinlari Is-Türk Ltd. Sti.
Atatürk Bulvari No. 191/Kat 13
Kavaklidere/Ankara Tel. 428.11.40 Ext. 2458
Dolmabahce Cad. No. 29
Besiktas/Istanbul Tel. (312) 260 7188
Telex: (312) 418 29 46

UNITED KINGDOM – ROYAUME-UNI
HMSO
Gen. enquiries Tel. (171) 873 8496
Postal orders only:
P.O. Box 276, London SW8 5DT
Personal Callers HMSO Bookshop
49 High Holborn, London WC1V 6HB
Telefax: (171) 873 8416
Branches at: Belfast, Birmingham, Bristol, Edinburgh, Manchester

UNITED STATES – ÉTATS-UNIS
OECD Publications and Information Center
2001 L Street N.W., Suite 650
Washington, D.C. 20036-4910 Tel. (202) 785.6323
Telefax: (202) 785.0350

VENEZUELA
Libreria del Este
Avda F. Miranda 52, Aptdo. 60337
Edificio Galipán
Caracas 106 Tel. 951.1705/951.2307/951.1297
Telegram: Libreste Caracas

Subscription to OECD periodicals may also be placed through main subscription agencies.

Les abonnements aux publications périodiques de l'OCDE peuvent être souscrits auprès des principales agences d'abonnement.

Orders and inquiries from countries where Distributors have not yet been appointed should be sent to: OECD Publications Service, 2 rue André-Pascal, 75775 Paris Cedex 16, France.

Les commandes provenant de pays où l'OCDE n'a pas encore désigné de distributeur peuvent être adressées à : OCDE, Service des Publications, 2, rue André-Pascal, 75775 Paris Cedex 16, France.

7-1995

OECD PUBLICATIONS, 2 rue André-Pascal, 75775 PARIS CEDEX 16
PRINTED IN FRANCE
(66 95 19 1) ISBN 92-64-14590-7 - No. 48234 1995